CONTENTS

CONTENTS

Chapter 1

The Sustainable Development Goals

In 2015, the United Nations created the 17 *Sustainable Development Goals* - 17 distinct goals to be met by the year 2030 in order to ensure "peace and prosperity for people and the planet, now and into the future," Each of the 17 goals covers a distinct humanitarian, economic, environmental, or social concern that the UN believes ought to be targeted to ensure peace and prosperity. 17 goals may not sound like a lot to create a utopian society that lasts for generations, but each of the 17 goals contains dozens of more specific *targets* and *indicators* (169 and 232, respectively) that the UN hopes we, as a species, may be able to achieve.

See, for example, Goal 3. The official wording of the goal is "to ensure healthy lives and promote well-being for all at all ages." This might sound a bit broad - and that's because it is: In the heart of this goal lies specific targets - reducing maternal mortality, reducing road injuries and deaths, promoting mental health, etc. - that we can look at to see if the goal has truly been met.

With important—and ambitious—goals set out by the UN and actionable steps in place to achieve these goals by 2030, you might guess that significant progress has been made towards achieving these Sustainable Development Goals. You may even be brave enough to hope that, nine years after the goals were set in place, at least *one* or *two* of them have been completed.

THE GOALS

The United Nations' 17 Sustainable Development Goals	**1. NO POVERTY**
2. ZERO HUNGER	**3. GOOD HEALTH AND WELL-BEING**
4. QUALITY EDUCATION	**5. GENDER EQUALITY**
6. CLEAN WATER AND SANITATION	**7. AFFORDABLE AND CLEAN ENERGY**
8. DECENT WORK AND ECONOMIC GROWTH	**9. INDUSTRY, INNOVATION AND INFRASTRUCTURE**
10. REDUCED INEQUALITIES	**11. SUSTAINABLE CITIES AND COMMUNITIES**
12. RESPONSIBLE CONSUMPTION AND PRODUCTION	**13. CLIMATE ACTION**
14. LIFE BELOW WATER	**15. LIFE ON LAND**
16. PEACE, JUSTICE AND STRONG INSTITUTIONS	**17. PARTNERSHIP FOR THE GOALS**

Well, dear reader, you would be wrong. In fact, an astonishing *zero* of the goals have been met today, and a flabbergasting *zero* are likely to be met by 2030. To be fair - the COVID-19 pandemic did exacerbate existing issues, undoubtedly impacting every single goal on the list to some extent. Even before COVID-19, however, we weren't on track to meeting the goals' ambitious targets.

Part of the reason for this may be the goals themselves. 17 goals with a total of 169 targets is a *lot* of goals. Still, the UN is a powerful, sufficiently funded entity with the support of governments worldwide. Another reason may lie in the individual goals, which often seem to contradict themselves and each other: on first glance, how could we possibly balance economic growth with environmental sustainability, and balance environmental sustainability with ending world hunger (though these "contradictions" will both be addressed throughout this book!)

So, what is the problem? Billions of dollars have been spent by non-governmental organizations (NGOs, for short) and government entities to solve these issues, which are of *immense* importance to ensuring the prolonged prosperity of our species and our planet. Yet, still, it seems like we can't buy an inch of progress.

The root of the problem lies in the internal controversies that develop on the way to solutions. Everyone agrees on Sustainable Development Goal 1, target 1: *Eradicate extreme poverty for all people everywhere, currently measured as people living on less than $2.15 a day*. Although this is undoubtedly a noble goal, how can we as a society work to fix this goal, when almost all potential solutions are controversial? Particularly in an interconnected yet *divided* world, any time a bill that can make real change is authored, there may very well be a person arguing that the bill is too costly for a country's taxpayers - and that person may be right! When you consider the important factor that the Sustainable Development Goals are looking for solutions to global problems, the possibility of a "good" solution becomes even less likely to remain uncontroversial due to the great amount of political and cultural diversity in the world in which we all live.

"Nothing comes to my desk that is perfectly solvable... Otherwise, someone else would have solved it. So you wind up dealing with probabilities. Any given decision you make you'll wind up with a 30 to 40 percent chance that it isn't going to work." - President Barack Obama

I recently heard from an activist working with an NGO in the United States about her important work to prevent teen pregnancies and STDs in her community. Her organization observed that, in CVS drug stores, condoms were locked behind boxes such that a person wanting to purchase a condom would need to speak to a staff member so that the staff member could unlock the box and sell the condoms. The organization believed that this made teenagers in the community less likely to protect themselves during sex, raising the number of STDs and teen pregnancies in their area. In order to test their theory, members of the organization *dressed up as teenagers*, walked into CVS stores, and attempted to purchase condoms. They then documented their difficulties in doing so, showing that, for a real teenager, purchasing a condom would be extremely difficult.

Members of the organization believed that the issue was contributing to the spread of STDs so negatively that they decided to contact CVS's corporate division in order to get condoms out of the locked compartments. After months of campaigning, CVS listened! The condoms were out of their locked compartments, and teenagers could grab them and proceed to self-checkout without having an awkward or judgemental interaction with employees. After this change, the organization observed a drop in the number of teen pregnancies and STDs in their area.

The first thing that everyone reading this should take from this story is inspiration - the organization did something that, to be honest, anyone could do (even you!) and caused *measurable and real change* in their community. The main point of the story is, however, that *even though* the organizations attacked the root cause of the issue and advocated for policies that resulted in measurable change, there are still arguments to be made against their work. For example - what if making condoms more accessible incentivized teenagers who would not have had sex without condoms to have sex at an earlier age? When I asked the activist this, she explained that the issue with STDs and teen pregnancies had gotten so bad in her community that the benefits outweighed any potential harms - understandable. Other arguments against her position might be made to - what if making the condoms more accessible caused more shoplifting? What if making condoms more accessible led to teenagers misunderstanding safe sex practices and relying solely on condoms without proper education on their effective use and other options for contraceptives?

Every government policy, no matter how seemingly wonderfully positive it may be for society, has downsides and arguments like this. Although making condoms had a positive local effect, one could (and one would, on an international scale) make arguments that would disparage the organization's efforts.

Additionally, each and every issue being discussed in public policy - no matter how simple it may seem - must be considered from all angles and perspectives, as almost everything in society is interconnected. Take the Sustainable Development Goals as an example: although the goals cover a diverse group of topics and issues that are pervasive throughout societies around the world, the goals are all interconnected. For example, see goal 3, good health and well-being, and goal 10, reduced inequalities: although the uninformed reader might argue that these two goals are entirely separate from each other, the issues are more closely related than that uninformed reader might think. Take just the first target in goal 3 - reduce global maternal mortality rates - one would think that the United States, with its technological advancements and vast wealth, would be leading the world in this metric. However, the United States actually has a far higher maternal mortality rate than every other high-income country. Although one might point to the lack of a universal healthcare system in the United States as part of the reason for this disparity, statistics show that the real reason for the United States' high maternal mortality rate is its *racial* inequalities. Specifically, the maternal mortality rate for Black women is 2.6x higher than the maternal mortality rate for White women in the United States. This stems from a variety of reasons - due to systemic inequities, Black women are more likely to live in underprivileged areas with limited healthcare access and often receive less care in hospitals due to unconscious bias and stereotypes. In order to support the good health and well-being of all women in America and across the globe, system inequalities *must* be reduced. This shows that the Sustainable Development Goals - whether discussing good health or inequality - are *all* related. Thus, whenever we want to make a policy that seems beneficial in one arena (economic, environmental, social, ethical, etc.), we need to consider the other arenas we are affecting. You will be hard-pressed to find any public policy solution that benefits every single arena - there will always be something for the devil's advocate!

This leads us to the situation we are in today - supplied with outcomes that we, as a society, know we want, but no sufficiently utilitarian method exists in order to achieve these outcomes. Even *if* everyone was well-educated on the topics at hand and came from a similar perspective, uncontroversial and universally beneficial methods would *still* not even exist. All methods of solving any problem with real depth have drawbacks. So how can we, as a society, even do anything? How can we balance and objectively view a policy when we know there will be both pros and cons? This is something that will be explored in depth throughout the remainder of this book - specifically, how can we create policies that provide real, tangible benefits for society while being economically and environmentally responsible and sustainable? How can we leverage both quantitative evaluations and qualitative evidence in order to pick what policy is "the best" on a global scale?

Throughout the rest of this book, we will explore four of this decade's biggest issues pertaining to the subject of global health that explicitly relate to the UN's Sustainable Development Goals. Additionally, the issues selected are all underreported relative to their importance to global health at large. Each section, which will stand alone (and work together) in a way analogous to the goals themselves, will explain all sides of the issue while eventually providing an analysis of all possible solutions.

Importantly, this book does not contain original quantitative research or evaluations. We won't be employing t-tests in order to analyze data and determine whether or not a given policy is successful (so it's okay if you don't even know what a t-test is!) - instead, we will be using qualitative methods by examining social, political, and economic trends over time to get a picture of what's happening *beyond* the data. Still, sources employed do take use of quantitative modeling and large-scale studies - so that sort of quantitative analysis is embedded within the lines of each chapter.

I hope that the specific issues selected and solutions discussed can serve as examples of ways that we, as a society, can approach targets listed by the UN in their Sustainable Development Goals. While the specific issues discussed will not be as broad as any of the 17 Sustainable Development Goals themselves (and, in some cases, be more specific than any individual indicators), the breadth and depth of the issues will encompass multitudes of goals. Through discussing GMOs, we will touch on important issues such as world hunger (Goal 2!) and economic growth (Goal 12!). Through an introduction to Antibiotic Resistance, we will examine global health (Goal 3!) and possible avenues for innovation (Goal 9!). Importantly, each chapter will spend *significantly* more time explaining and researching each issue than discussing potential solutions. This is meant to emphasize the importance of good research in creating and analyzing solutions to global problems - after spending so much time looking at an issue from every possible orientation, you will most likely find the solutions discussed to be borderline *obvious*. You may even have your own ideas for solutions that stem from reading the research that the book provides.

Thus, the utility of this book is threefold: firstly and primarily, it serves as an example of how solutions to an issue can be effectively created and analyzed by utilizing years of existing scientific research and then applying existing scientific innovations, economic analysis, and social principles to the wealth of scientific research publicly available. Additionally, this book serves as a call to action - both action in terms of global progress to address the issues discussed herein, but also action in terms of creating better policy that is equitable, culturally understanding, and actionable. The second chapter will discuss antibiotic resistance and show how social, governmental, and scientific objectives can converge to create solutions to complex problems. The third chapter will discuss HIV/AIDS, which will show how scientific and medical factors can be exacerbated by social factors. The fourth chapter will discuss genetically modified organisms to show the reverse - how social factors can exacerbate existing issues within global health. The fifth and final chapter will discuss smoking to emphasize the interconnectedness of social factors within the study of global health, and will clearly consider distinct possibilities for solutions. Chapter six will highlight the subject of global health itself and include key takeaways.

Secondly, the explorations of the specific issues embedded throughout this book create solutions that could be useful to NGOs or government organizations when considering how to tackle these issues, which are all highly relevant to society on a large scale today. Additionally, this book hopes to draw necessary attention to the underrepresented issues discussed and the Sustainable Development Goals overall. Many issues are unnecessarily stigmatized or underreported, resulting in a lack of global progress on solvable issues. Thirdly, many of the global health issues discussed contain themes and specific issues that are relevant to a person's everyday life. By reading an essay, you will gain the knowledge necessary to make an informed choice when you are considering whether to smoke e-cigarettes or buy GMO food. This book will provide clear and concise recommendations, demystifying politically contentious subjects like HIV and smoking: how can you avoid being infected with HIV and mitigate its harms if you do? Are vapes and e-cigarettes as dangerous as traditional cigarettes, and do they emit harmful second-hand smoke? Are GMOs safe to consume on a daily basis? What can you do to heal from viral infections whilst not creating antibiotic-resistant viruses?

For the answers to these questions and more, continue on.

Chapter 2

Antibiotic Resistance and its "Solutions"

ANTIBIOTIC RESISTANCE: THE PROBLEM

On the heels of a worldwide pandemic looms another crisis that you *may not even have heard of* and offers one of the greatest global health challenges of the 21st century and challenges the very foundations of modern medicine. In order to understand why antibiotic resistance is so important, we need first to understand what antibiotics are. Antibiotics are medicines used to prevent and treat infections caused by bacteria. Bacteria are single-celled organisms that do *not* possess membrane-bound nuclei, making them prokaryotes. Although most bacteria found within the confines of a human body are beneficial and even necessary for human survival, some types of bacteria cause infections like strep throat, E. coli, and urinary tract infections. These infections are thus treatable by antibiotics. There are two different types of antibiotics: bactericidal antibiotics and bacteriostatic antibiotics. Bactericidal antibiotics work by killing the bacteria that they are targeting entirely. Bactericidal antibiotics are usually able to do this by interfering with specific elements of bacterial biology, such as the formation of the cell wall. Penicillin, the first antibiotic, is one example of a bactericidal antibiotic.

Contrasting bactericidal antibiotics are bacteriostatic antibiotics, which work by stopping bacteria from reproducing. Similar to bactericidal antibiotics, bacteriostatic antibiotics work by interfering with specific elements of bacterial biology. However, they most commonly function by hindering the bacteria's ability to synthesize proteins as opposed to the bacteria's ability to form the cell wall.

As explained, both bactericidal and bacteriostatic antibiotics work by targeting specific aspects of bacterial biology. Importantly, these mechanisms do *not* exist in viruses because viruses are *not* complete cells; they are merely genetic material such as DNA and RNA surrounded by a protein coating that needs a cell to survive. Since viruses don't have the distinct biology that antibiotics target in bacteria, antibiotics are entirely ineffective against viruses. Not only is using antibiotics to treat viral infections ineffective, but it's also dangerous; taking an antibiotic for no reason unnecessarily opens a person up to various negative side effects of antibiotics (most commonly digestive problems or headaches). Further, misuse and overuse of antibiotics are key factors that have recently led to a worldwide increase in antibiotic resistance.

Antibiotic resistance occurs when bacteria develop resistance mechanisms in order to better defend themselves from the effects of antibiotics. Bacteria which are resistant to antibiotics are more dangerous to humans as they are less easily treatable. While a majority of people surveyed by the World Health Organization believe that antibiotic resistance means that a person's body no longer responds to antibiotics, this is a common misconception. In reality, the bacteria themselves become resistant to antibiotics, most often due to a person misusing antibiotics. In order to become resistant to antibiotics, bacteria develop *resistance mechanisms*. When an already hard-to-treat bacteria develops various resistance mechanisms, treatment becomes almost impossible, and the bacteria become resistant to antibiotics. While some resistance mechanisms change the bacteria, many resistance mechanisms involve changing or destroying the antibiotic. For example, the bacteria *Klebsiella pneumoniae* produces enzymes called *carbapenemases*, which break down antibiotics, rendering them useless. Other resistance mechanisms involve bypassing the effects of the antibiotic and restricting the antibiotic's ability to enter the bacteria through the bacteria's cell wall.

Bacteria develop various resistance mechanisms by a process known as natural selection. In biology, natural selection is a naturally occurring process that results in the adaptation of an organism to its environment over time. Since organisms that are best suited to their environments have the best chance to reproduce and spread their genes, these genes eventually spread throughout a population, often leading to evolution. When a bacterium reproduces, there is a small chance of a mutation that could result in the bacterium's offspring developing resistance mechanisms. If the entire bacteria population is exposed to an antibiotic, those without restriction mechanisms will die, leaving the small number of bacteria that naturally mutated resistant mechanisms to spread their genes to the succeeding generations. Thus, the introduction of antibiotics acts as a *selection pressure* that, while killing some bacteria in the short term, actually strengthens the bacteria's ability to resist the effects of antibiotics in the long term. This isn't to say that all antibiotics are bad, however. Although this chapter might come across as 'anti-antibiotics,' antibiotics are overall extremely positive for global health. Antibiotics have saved millions of lives each year, so much so that they alone increase life expectancy in the United States by almost 10 years. The real problem - and the *solvable* problem - is the misuse and overuse of antibiotics. When people use antibiotics that are not prescribed to them or use antibiotics for a shorter or longer time than they are prescribed, they significantly increase the chances of antibiotics developing resistance mechanisms.

Not only does misuse and overuse affect the ability of bacteria to develop resistance mechanisms, but it also strengthens these resistance mechanisms and other factors that lead to the bacteria becoming more dangerous ("dangerous" both in terms of how infectious the bacteria is and in terms of how the bacteria can negatively affect human health). Over time, as the entire bacteria population naturally evolves to develop antibiotic resistance mechanisms, the mechanisms become stronger each time the bacteria is exposed to antibiotics since the bacterium with the best resistance mechanisms is the most likely to survive and pass on its genes. The fact that bacteria are often extremely numerous and reproduce rapidly makes the odds of a colony of bacteria developing some sort of resistance mechanism high. In addition to natural selection's effect on antibiotic resistance, natural selection also plays a role in the virulence and pathogenicity of infectious agents like bacteria or viruses. Over time, infectious agents with greater virulence may outcompete less virulent strains, leading to the evolution of more harmful infectious agents. Similarly, more pathogenic bacteria or viruses are able to infect hosts more efficiently, allowing them to outcompete less pathogenic infectious agents. Therefore, due to natural selection, infectious agents will only continue to become even more virulent and pathogenic while developing resistance mechanisms that may one day render our antibiotics useless. The relatively new advent of antibiotics is adding selection pressure to this, meaning that infectious agents will become more virulent and pathogenic faster now than ever. Altogether, bacteria are becoming more virulent and pathogenic, while the antibiotics we have today are continuing to lose effectiveness as bacteria evolve to develop resistance mechanisms.

The current situation is dire: bacteria have already evolved resistance mechanisms to *every single* main class of antibiotics used today. The results of this development have been catastrophic: in the United States alone, antibiotic-resistant bacteria cause more than 2.8 million infections each year and at least 35,000 deaths. Outside of the United States, the problem is even worse: worldwide, antibiotic resistance is associated with nearly 5 million annual deaths. Deaths are the most common in Sub-Saharan Africa and South Asia, and the impacts of resistant infections have generally been the worst in low-income countries since these countries do not have access to the resources required to combat infections that are not easily treatable with antibiotics. The number of deaths associated with antibiotic resistance has concerningly increased in the last few years and is projected to keep increasing; one report projects resistant infections to cause *10 million deaths per year* and impose a global financial cost of *100 Trillion USD* by 2050. Many resistant infections currently have no treatment options, further illustrating the pressing need for a solution.

ANTIBIOTIC RESISTANCE: THE "SOLUTIONS"

While a bleak future with widespread antibiotic resistance and increasingly virulent infectious agents seems imminent, there are numerous solutions and strategies that have the potential to mitigate the crisis and preserve the efficacy of antibiotics for future generations.

There are many things that governments and hospitals can do in order to combat the antibiotic resistance problem in the United States. One of the most important things that governments and hospitals can do to curb antibiotic resistance is to reduce antibiotic use in human medicine. Even though all U.S. hospitals have some regulations on when antibiotics can be given, more than half of U.S. patients receive antibiotics during their stays, resulting in many (often unnecessary) opportunities for bacteria to evolve restriction mechanisms. Wider and more aggressive antibiotic stewardship programs, which aim to ensure appropriate antibiotic use and patient safety, are necessary to decrease the overall expansion of resistant bacteria. Current stewardship programs vastly differ in quality from hospital to hospital, with significant steps still necessary as there is a significant lack of a national or international stewardship program. On a national scale, current stewardship programs do not even have a definition of what "appropriate" antibiotic use even is. In order to foster better stewardship programs and create a global understanding of appropriate antibiotic use, governing bodies must establish empiric uses and misuses for antibiotics.

The largest governing body in the United States with regard to antibiotic stewardship is the Centers for Disease Control and Prevention's Core Elements for Antibiotic Stewardship. Although the CDC provides guidelines for hospitals and doctor's offices to abide by, these guidelines have significant issues. Firstly, they are not enforced whatsoever on a national scale. Secondly, the guidelines are often too vague and can thus be difficult for hospitals and doctor's offices in more rural areas that are potentially under-resourced to adjust to. Thirdly, many of the guidelines are unactionable and expensive; they provide advice that, while good in theory, doesn't provide steps for hospitals or doctor's offices to actually better their antibiotic stewardship programs without spending an unreasonable amount of money on local cooperation. A more centralized stewardship system - at least on the state level - that provides actionable, effective, and reasonable guidelines for all hospitals, pharmacies, and doctor's offices to abide by could vastly improve the United States' fight against antibiotic resistance, saving lives on a global scale. Stewardship systems in other countries could also benefit from these centralized and effective stewardship systems; however, solutions must *always* be catered to the specific needs of these countries and their existing medical infrastructure.

Even though antibiotic stewardship programs have improved significantly throughout the 21st century, the antibiotic resistance crisis has yet to turn a tide - partially due to human's collective overuse of antibiotics in other ways, such as feeding antibiotics to animals. Currently, the agricultural sector's use of antibiotics on animals is entirely prolific (13.5 million pounds of antibiotics for animals per year), even though antibiotic use in agriculture has been recognized as a contributor to the international growth of antibiotic resistance. There are three distinct, simple, and actionable ways for the government to help curb the agricultural sector's use of antibiotics on animals. For one, the government should strengthen the regulation of feeding farm animals antibiotics; even though there is currently some regulation on antibiotics in farming, the Environmental Protection Agency could set more rules in accordance with how they already regulate the usage of certain chemicals in agriculture or align their rules for providing antibiotics to animals to their rules for providing antibiotics to humans. A way that the government could directly involve consumers in the fight against antibiotic resistance is by creating a special label called "No Feed-Antibiotics." Consumers would then know which products are antibiotic-free and would potentially view these products as either more healthy or more environmentally conscious. Over time, this would result in a market incentive for farmers to use less and less antibiotics. Lastly, all medically important antibiotics for use by animals should have to be brought under veterinary oversight, as opposed to letting farmers purchase over-the-counter antibiotics for their animals, as is often the current system. Potential downsides of this policy are twofold: firstly, antibiotic sales would be significantly lowered, hurting the revenue of drug makers and potentially contributing to (small) negative economic effects. Secondly, there could be difficulties in the short term in establishing a method for animals who need antibiotics to get them, and some animals may go untreated. However, over time, a more robust system allowing farm animals to get prescriptions could eventually be put in place with the help of the federal and state governments that would alleviate this issue.

Alternatively to the aforementioned solutions, a final action that the United States government could consider taking to combat antibiotic resistance is *lifting* regulations on corporations that produce and design antibiotics. While regulation on how antibiotics are used is certainly appropriate and necessary, there has recently been a lack of innovation by corporations due to extremely strict government regulations on antibiotic production. While it may seem as though more antibiotics would be a bad thing in the fight against antibiotic resistance, new innovative antibiotics are a necessity because they can adapt to bacteria's more recent evolutions and be able to kill or stop the reproduction of bacteria that are currently considered "antibiotic resistant" and even incurable. Balancing regulation that can effectively curb the growth of antibiotic resistance while allowing for more antibiotic innovation may sound like an impossible endeavor, but nuanced and fact-based solutions *can* exist: lifting regulations on medical companies can, counterintuitively, go hand-in-hand with increasing regulations on farms and hospitals. However, like always when discussing issues on this large a scale, economic, environmental, and social factors must be considered to ensure that all enacted public policies provide a net benefit to society.

The aforementioned methods to combat antibiotic resistance might make most people feel powerless, but we can all play a part in preventing resistance. The *worst* thing we can do is to "save" antibiotics so we can take them later, when we weren't prescribed them, and take antibiotics that were prescribed for someone else. Additionally, we should all take antibiotics exactly as prescribed, not skipping doses even if we are feeling 100%

better. Lastly, we can all maintain good hygiene in order to control infections.

It is certainly within our capability to slow down the growth of antibiotic resistance, as has been shown. However, bacterial evolution is ancient, active, and continuous; it is impossible for us to completely "solve" the antibiotic resistance crisis forever since we can not stop bacteria from evolving, and it is impossible for us to completely eradicate bacteria. All we can do is continue to innovate. If we do so, our antibiotics will continue to become better and will outpace bacteria's rate of evolution, making the issue moot. However, this is not as easy as it sounds; currently, just 27 new antibiotics are in the clinical trial stage, with just six innovative enough to overcome antibiotic resistance. Humanity's biggest priority should be increased antibiotic innovation, which will only be achieved when it is both financially sensible and legally feasible for corporations to do so. In addition to cutting back legal restrictions, as has been discussed, governments should provide economic stimulus to innovative corporations and support early investment into research. Urgent and concerted investments from governments, corporations, and private entities alike are "needed" in order to create a viable system for antibiotics. This economic and legal support would give antibiotic innovation the shove it needs in order to push out new antibiotics faster than bacteria can evolve, making antibiotic resistance an irrelevant issue since there would always be a new antibiotic coming for "resistant" bacteria.

There have been various notable ways that scientists today are innovating to "solve" the antibiotic resistance issue. One group of researchers focused on a protein called DsbA, which is found in bacteria and plays a crucial role in the folding of proteins that help bacteria create resistance mechanisms (which, themselves, are often proteins). The researchers found a way to inhibit the DsbA protein in bacteria, therefore weakening the bacteria's ability to resist antibiotics. This has the potential to neutralize most bacteria's resistance mechanisms and, in combination with other antibiotics, cure previously incurable diseases. In addition to this innovative solution, scientists have worked on making improvements to existing drugs in order to combat resistance. One such drug, linezolid, is used against bacteria such as streptococci and staphylococci, which cause pneumonia and pus formations, respectively. However, many streptococci and staphylococci strands have now become resistant to linezolid. Scientists are working on manufacturing tedizolid and radezolid, two new modifications of linezolid that offer improvements over the traditional version of the drug, both because they work on linezolid-resistant strains and because they can work on a wider range of bacteria. This is a huge achievement in the fight against linezolid-resistant bacteria, but there needs to be progress like this for every strand of antibiotic-resistant bacteria that is harmful to humans - of which there are at least 18 that have caused deaths in the United States and 9 that the WHO lists as of critical or high priority.

Various antibiotics have been approved throughout 2023, both for the purpose of combatting antibiotic resistance and for combatting previously untreatable antibiotics. One antibiotic that has been used to combat antibiotic resistance is called Xacduro. This drug looks to specifically challenge the bacteria Acinetobacter baumannii, which causes bacterial pneumonia. Acinetobacter baumanni is particularly resistant, and Xacduro is notable since it is made exclusively for Acinetobacter baumannii, while most other drugs are meant to work for multiple organisms. The antibiotic is anticipated to be available later this year. Antibiotics like Xacduro are key to solving the antibiotic resistance crisis. While Xacduro is the only antibiotic to have been

FDA-approved in 2023, there are other drugs coming down the pipeline, including a particularly interesting antibiotic called Clovibactin. Clovibactin was isolated from unculturable bacteria and is able to combat harmful bacteria, including those with numerous resistance mechanisms. Additionally, bacteria treated with Clovibactin did not develop any detectable resistance (yet). While still in its infancy, Clovibactin and antibiotics like it could be key to substantial innovation and even a long-term "solution" to the antibiotic resistance problem. Although there is still a pressing need for more antibiotic innovation, Xacduro and Clovibactin are proof that such innovation can result in life-saving antibiotics.

Artificial Intelligence has been used by scientists to both discover and screen antibiotics in order to save lives and combat the antibiotic resistance crisis. Using a machine-learning algorithm, researchers have leveraged Artificial Intelligence to screen hundreds of millions of chemical compounds in order to find potential antibiotics that kill bacteria in new ways. Since the drugs that Artificial Intelligence screens are fundamentally unlike the antibiotics available today, antibiotic-resistant bacteria often have no resistance mechanisms to safeguard them, meaning Artificial Intelligence's drugs are able to combat bacteria that were previously thought to be resistant in huge numbers and with high success rates. Drugs discovered by Artificial Intelligence have been put to the test in recent years, with one such drug being able to fend off Acinetobacter baumanni (the same bacteria that Xacduro targets) in mice. Antibiotics created by Artificial Intelligence have a long way to go before reaching the clinic because their novel methods require extensive testing to ensure they will be safe for human use. However, the drugs are extremely promising; the one meant to fend off Acinetobacter baumanni, for example, produced successful results on all 41 strands of the bacteria that it was tested on. Drugs procured by Artificial Intelligence provide a novel solution to the antibiotic resistance problem because its drugs are unique from anything humans have created, meaning bacteria have not had the opportunity to evolve to resist them. While this innovation is important and will likely save lives, it is, again, not a permanent solution for the antibiotic resistance crisis, as bacteria will simply evolve new resistance mechanisms to combat these new antibiotics.

In conclusion, the antibiotic resistance crisis presents humanity with a significant global health challenge that threatens the foundations of modern medicine. Driven by the effect of natural selection on bacteria, this issue is complex and will never fully be solved. However, there are numerous things that we can do to combat the crisis, including reducing antibiotic usage for both humans and animals and, most importantly, fostering innovation in antibiotic development by supporting medical companies legally and financially. Successful medications like Xacduro have shown that such innovation will pay off in the battle against antibiotic resistance, while the promise of Artificial Intelligence is even more evidence that human innovation will help to mitigate the antibiotic resistance crisis. Throughout the methods described herein, it can be concluded that it is possible to preserve the efficacy of antibiotics and protect the health of future generations by fostering increased innovation.

Chapter 3

The Road to Immunity: A Comprehensive Exploration of HIV

HUMAN IMMUNODEFICIENCY VIRUS: HISTORY & GLOBAL IMPACT

Human immunodeficiency virus, or HIV, is a virus that damages the human body's immune system, interfering with its ability to fight infection and disease. When untreated, HIV can lead to acquired immunodeficiency syndrome, or AIDS, which has killed approximately 40.4 million people since the start of the epidemic. This chapter will thoroughly explain the HIV epidemic and its significance, including how HIV is treated and prevented. Additionally, this chapter will discuss the history of HIV/AIDS, immunity to HIV, and the implications of gene editing. Altogether, this chapter will emphasize the importance of commitment to putting an end to the suffering and global disparities that HIV has caused.

As of 2023, HIV/AIDS is a globally recognized epidemic, with December 1st being World AIDS Day and 39 million people living with HIV. Due to the number of people impacted by HIV/AIDS, therapies and drugs have been developed in order to combat HIV/AIDS, with 29.8 million, or 76%, of the 39 million people living with HIV having access to antiretroviral therapy (the primary treatment to manage HIV). HIV is also continuing to spread to broad populations, with cases all around the world. However, the infection is spreading less than it has in the past; in 1995, as many as 4.3 million people became newly infected with HIV/AIDS, compared to 1.3 million people in 2022. The infection's negative effects on humans range beyond just health effects. For example, poverty has been shown to be higher among households affected by HIV/AIDS as compared to unaffected households. HIV/AIDS has also been shown to further income loss for people already in poverty, affecting their ability to meet basic needs. Additionally, there is a significant amount of stigma surrounding HIV and AIDS, particularly in Africa and Asia, where HIV and AIDS have been shown to contribute to a higher risk of mental health disorders and issues such as anxiety (due not to the virus and disease itself, but to social factors).

While the HIV epidemic is often thought of as beginning in 1981 (since that is when the disease was discovered in the United States), the history of the virus traces back to Central Africa in the early 19th century. At that time, chimpanzees in West Africa were infected with simian immunodeficiency virus, a virus similar to HIV that attacks the immune system of monkeys and apes. One particular strain of simian immunodeficiency virus, known as SIVcpz, is nearly identical to the HIV seen in humans. SIVcpz likely jumped to humans when hunters in Africa ate infected chimps. This is believed to have led to a global pandemic in 1920 in Kinshasa, the largest city in the Democratic Republic of Congo. The spread of HIV was facilitated by international travel as it continued to spread around Africa and eventually was discovered in the United States. In the United States, the disease was "discovered" in 1981 and named Gay-Related Immune Deficiency because it was believed to only affect men. As scientists found that the disease could affect heterosexual individuals and even women, the disease was renamed Acquired Immune Deficiency Syndrome, or AIDS. In 1982, scientists realized that AIDS was sexually transmitted and caused by HIV. By 1987, the first drug was approved to treat HIV/AIDS.

HUMAN IMMUNODEFICIENCY VIRUS: TRANSMISSION, PREVENTION, & TREATMENT

Like all viruses, HIV requires a host cell to replicate. Unlike most other viruses, however, HIV is a retrovirus. Retroviruses are a unique type of virus characterized by their RNA (ribonucleic acid) genomes and their ability to convert their RNA genomes into DNA (deoxyribonucleic acid) with an enzyme called reverse transcriptase during a process known as reverse transcription. Retroviruses' unique ability to convert their RNA genomes into DNA is used when they infect host cells. After retroviruses infect a host cell, they use the enzyme reverse transcriptase in order to convert their own RNA into DNA, which is then integrated into the host cell's DNA, thus changing and infecting the genome of the cell. This process is unique and important because it allows the retrovirus to become a permanent part of the genetic material of the host cell. After the retrovirus fuses with a host cell, it replicates itself, and the new copies of the retrovirus leave the host cell in order to infect other cells. Since HIV is a retrovirus, this is how HIV infects host cells. HIV is special because it targets CD4 + T cells in the immune system. These CD4 cells are vital for fighting off infections and allow HIV to reach a variety of environments within the human body.

Approximately 40% of new HIV infections are transmitted by people who did not even know that they had the virus. Additionally, the average lifespan for someone who develops AIDS is only three years if left untreated. These figures illustrate the dire need for accurate and frequent testing. There are three types of HIV tests, which are all considered accurate. However, the tests can not detect HIV immediately after infection because there will be very few viruses in an individual's bloodstream right after they are infected. The first method of HIV testing is antibody tests. An antibody is a protein component in an individual's immune system that circulates in the bloodstream, recognizing and neutralizing harmful viruses and bacteria. Antibodies are naturally produced by an individual's immune system when exposed to a virus. Antibody tests look for the antibodies specific to HIV in a person's blood or oral fluid. Antibody tests are often the only test accessible to many in the United States because the only FDA-approved HIV self-test and most rapid tests are antibody tests. The second method of HIV testing is antigen/antibody tests. Antigens are foreign substances produced by harmful bacteria or viruses that cause a person's immune system to activate. Since antigens are produced before antibodies have a chance to develop, antigen/antibody tests can determine if a person has HIV sooner after the initial contraction of HIV than antibody tests. In the case of HIV, the antigen developed is called p24. Antigen/antibody tests are the most common in the United States. The third method of HIV testing is nucleic acid tests. Nucleic acid tests look for HIV itself in the human bloodstream. A nucleic acid test can be more accurate than the other two testing methods for this reason. Additionally, a nucleic acid test can detect strands of HIV in an individual's bloodstream earlier than antibody or antigen/antibody tests since the strands of HIV will be present before antibodies or antigens develop.

When an individual reaches the advanced stage of an HIV infection, they are diagnosed with AIDS, the final stage of the infection caused by HIV. Specifically, AIDS is diagnosed when an individual with HIV experiences one of a few opportunistic infections or cancers associated with HIV or has an extremely low number of CD4 cells (below 200 cells per cubic millimeter of blood). While many individuals with HIV do not even exhibit any symptoms, the same is not true for AIDS. AIDS is characterized by symptoms including weight loss, diarrhea, fever, and night sweats. Without treatment, an individual with AIDS has a life expectancy of three years. This life expectancy falls to just one year if a person is diagnosed with an opportunistic infection associated with HIV. Importantly, every HIV infection does not necessarily progress to AIDS: proper medical care can significantly slow the progression and allow individuals with HIV/AIDS to lead healthy lives.

HIV is primarily treated with antiretroviral therapy, a combination of medications that target different stages of the HIV life cycle with the goal of slowing the progression of the virus and managing the symptoms. Importantly, HIV currently has no cure: antiretroviral therapy serves solely to mitigate and manage the symptoms and progression of HIV. Access to proper antiretroviral therapies can significantly impact a person with HIV's quality of life and lifespan and can prevent a person's HIV from progressing to AIDS. There are six main classes of antiretroviral drugs that are used to treat HIV. While each class attacks the virus in a different way, the standard approach to HIV treatment is to use multiple drugs from different classes in combination.

The first class of antiretroviral drugs is nucleotide reverse transcriptase inhibitors. As the name suggests, these drugs work by targeting the protein reverse transcriptase. Specifically, they disrupt the construction of the new piece of DNA, in doing so, halting HIV replication. Nucleotide reverse transcriptase inhibitors are often referred to as the backbone of HIV treatment. Such drugs as Abacavir, Emtricitabine, Lamivudine, Tenofovir disoproxil, Tenofovir alafenamide, and Zidovudine all fall under the category of nucleotide reverse transcriptase inhibitors.

The second class of antiretroviral drugs is non-nucleoside reverse transcriptase inhibitors. They also target the enzyme reverse transcriptase, but do so by binding directly to the enzyme, as opposed to doing so by disrupting the construction of a new piece of DNA like nucleotide reverse transcriptase inhibitors do. Such drugs as Doravirine, Efavirenz, Etravirine, Nevirapine, and Rilpivirine all fall under the category of non-nucleoside reverse transcriptase inhibitors.

The third class of antiretroviral drugs is integrase inhibitors. Integrase inhibitors target the protein integrase, which is essential for the viral replication of HIV. Integrase binds to the host cell targeted by HIV before preparing the area for integration and transferring the processed HIV strand into the host cell's genome. Therefore, integrase inhibitors are able to stop HIV from integrating itself into the DNA of human cells. Such drugs as Bictegravir, Dolutegravir, Elvitegravir, Raltegravir, and Cabotegravir all fall under the category of integrase inhibitors.

The fourth class of antiretroviral drugs is entry inhibitors. Entry inhibitors prevent HIV from entering human cells. There are two types of entry inhibitors: fusion inhibitors and CCR5 inhibitors. A fusion inhibitor is usually reserved for people without other treatment options. The only FDA-approved fusion inhibitor is known as Enfuvirtide, which works by attempting to entirely stop the fusion of the HIV protein with the target CD4 cell. It is not commonly used because of its need for twice-daily injections and its high rate of injection site reactions. CCR5 inhibitors are more commonly used; they block HIV from attaching to one of the host cell's receptors, the CCR5 co-receptor. Similarly to fusion inhibitors, CCR5 inhibitors are very rarely used. In order to take CCR5 inhibitors, individuals with HIV have to take a test to see if it would be effective for them. The only approved CCR5 inhibitor is Maraviroc.

The fifth class of antiretroviral drugs is protease inhibitors. Protease inhibitors work against HIV by inhibiting the enzyme protease. Protease is used by HIV to break up large proteins and assemble new viruses. Protease inhibitors do not entirely stop HIV from replicating, but the replicated HIV viruses are immature and unable to infect new cells after the introduction of protease inhibitors. Such drugs as Atazanavir, Darunavir, and Lopinavir all fall under the category of protease inhibitors.

The sixth and final class of antiretroviral drugs is post-attachment inhibitors. These drugs are sometimes classified as entry inhibitors and are generally newer. Post-attachment inhibitors work by preventing HIV proteins from engaging with co-receptors on the host cell after they have already engaged with the host cell. One example of a post-attachment inhibitor is Ibalizumab.

While the most common way for an individual to spread HIV to another individual is through anal sex, HIV can be spread in a variety of ways. HIV can be spread through vaginal sex, from a mother to a child (during either childbirth or breastfeeding), and through sharing needles, syringes, or other injection equipment. Outside of these avenues, the spread of HIV is uncommon. Various methods have been developed in order to prevent the spread of HIV. For example, the amount of HIV transmitted from mother to child has been significantly lowered in recent years since pregnant women are getting tested for HIV more often. If a pregnant woman has HIV, they can start HIV treatment and, in doing so, significantly lower the baby's chance of contracting HIV. Additionally, if a baby takes HIV medication for 4 to 6 weeks after birth, that baby's chances of getting HIV are less than 1%. Another method of preventing HIV is post-exposure prophylaxis, or

PEP. If an individual is inadvertently exposed to HIV, PEPs can effectively prevent them from contracting HIV with few side effects. Similar to post-exposure prophylaxis, pre-exposure prophylaxis, or PrEP, can reduce the chances of an individual getting infected with HIV if they know that they are going to be at risk of contracting HIV in the future. PEPs and PrEPs are highly effective. Contraction through drug injection accounts for approximately 1 in 10 HIV diagnoses in the United States. Somewhat obviously, the best way to prevent getting HIV from injection drug use is by avoiding injecting drugs. However, if an individual is in a circumstance where they cannot avoid using injection drugs, they can lower their chances of contracting HIV by taking PrEP or PEP and not sharing needles, syringes, or other drug equipment. The number one thing that individuals can do in order to prevent the spread of HIV is to practice safe sex. Choosing to abstain from anal or vaginal sex in favor of activities like oral sex mitigates the risk of HIV. If an individual is partaking in anal or vaginal sex, then condoms should always be used, as condoms are a highly effective way of preventing HIV and other sexually transmitted diseases. Additionally, taking PrEP as it is prescribed before having sex or taking PEP as it is prescribed after having sex can be an effective way of limiting an individual's chances of contracting HIV.

HUMAN IMMUNODEFICIENCY VIRUS: RESEARCH ADVANCES & CHALLENGES

Since HIV integrates itself into host DNA, it has been extremely difficult for scientists to create a vaccine. Live-attenuated vaccines, which use a weakened form of the virus that causes a disease in order to create a strong immune response so the body can fend off the virus when it is exposed to it in the future, would not work for HIV because there are concerns that the live attenuated virus would elicit disease. Additionally, there are numerous different "clades," or types of HIV. Creating a vaccine for one clade would likely be ineffective against others. mRNA vaccines, which are used to combat viruses such as COVID-19, create proteins in order to trigger an immune response that kills the virus if an individual comes in contact with it in the future. However, mRNA vaccines have been ineffective against HIV because HIV would require too many proteins that can be created in just one or even a few vaccines. Other innovative methods have all been ineffective against HIV - HIV's ability to disguise itself and take control of host cells has made it extremely difficult to cure. There have been more than 250 HIV vaccine trials, but no cure has been found. Even though there will be no vaccine in the next few years, scientists still have hope for a vaccine in the future, however. Additionally, preventative measures such as PrEP, PEP, and antiretroviral therapy have provided a bridge for the time until a vaccine is developed.

While HIV is a grave threat to global health, a small percentage of people have been shown to have some natural form of resistance to HIV. The main mechanism by which individuals resist HIV is a mutation of the gene that encodes CCR5, which is one of the host cell's receptors for HIV. This is not the only mechanism of HIV resistance, however. For example, it was recently discovered that a rare mutation of the protein TNPO3 also causes natural immunity to a clade of HIV. The various ways that people are naturally immune to HIV have intrigued scientists who are looking to find a cure for HIV. These scientists have looked at the ways that immune systems are able to naturally combat HIV in order to create new innovative ways that HIV can be prevented entirely or even cured for everyone.

In 2018, a Chinese scientist named Jiankui He claimed to have created the world's first gene-edited babies. He claimed to have used a gene editing technique known as CRISPR/Cas9 in order to target the gene that encodes CCR5, one of the host cell's receptors for HIV. He was inspired to target the gene that encodes CCR5 because, as previously explained, the main mechanism by which individuals naturally resist HIV is a mutation of this same gene. The system he used in order to perform "surgery" on the genes worked by using the Cas9 enzyme to precisely cut DNA, thereby deleting genetic information and altering the genes. In videos released by He's team, He explained that he used gene editing solely for the purpose of making the babies immune to HIV. He explained that his technology, while controversial for ethical reasons, enabled a father with HIV to bring his children into the world without fear of being discriminated against or having health issues due to having HIV (Although He used this argument for why he performed the procedure, in reality, during IVF, sperm is modified such that there would be no opportunity for the virus to be transmitted from the father). He also emphasized his disapproval of the idea of "designer babies" - babies who have their genes edited to be naturally smarter, better looking, more athletic, and have other "positive" qualities (The He Lab, 2018). While He received significant positive recognition, including being listed as one of the most influential people of 2019, He was largely condemned and was even imprisoned due to his actions. The ethical implications of using gene-editing technologies to alter human biology is something that has continued to be a subject of great debate. Additionally, He did not receive permission from any authority or his University to conduct his experiments. Authorities remarked that his work was unethical, crazy, and a huge blow to the reputation and development of Chinese science. Overall, He illustrated that gene editing techniques can be used in order to make individuals at least partially naturally resistant to HIV (the word "partially" is used because there is some evidence that some of the girls' cells did not have the ccr5 gene altered). While He's methods were unorthodox and unethical, He's experiment makes a future where immunity to HIV is common seem plausible. As biotechnology continues to advance, there may be other methods to make people immune to HIV with their consent.

From its origins in Central Africa to the present-day global epidemic, HIV has provided a significant global health challenge for humanity. The stark realities of the epidemic highlight the urgency for increased testing, awareness, research, and healthcare. The exploration of natural immunity and gene editing experiments have raised ethical concerns and simultaneously provided hope for potential breakthroughs in HIV prevention. Despite the absence of a cure, significant progress has been made in treatment, providing hope for those living with HIV/AIDS. HIV used to be a death sentence and is now something that millions of human beings live with, illustrating the importance and effectiveness of continued scientific progress.

Chapter 4

Insulin and Beyond: Exploring Genetically Modified Organisms

GENETICALLY MODIFIED ORGANISMS: DEVELOPMENT, BENEFITS, & CAPABILITIES

Genetically modified organisms, or GMOs, are plants, animals, or microbes that have had their genetic code tweaked by humans in some way, typically as the result of a high-tech genetic engineering process that attempts to alter their characteristics. Humans create GMOs for a variety of reasons - for example, humans have genetically modified some plants to make them resistant to viruses, insects, and pesticides. Additionally, GMOs have been used to create vaccines and are instrumental in the production of insulin, a life-saving medicine for people with diabetes. This chapter will explore the effects of GMO crops on society, discuss the importance and history of insulin production using GMOs, discuss GMO regulation and labeling, and discuss GMOs' ability to solve critical global issues in order to create a comprehensive analysis of the usage of GMOs in the 21st century.

In order to create a genetically modified organism, scientists use a technique known as gene editing, in which they make a tiny, controlled change in the DNA of a living organism. Gene editing involves carefully inserting a specific gene into the DNA of a single cell. Traditionally, a vast majority of the organism's genetic code remains unchanged during this process. Once the single cell has been modified, scientists stimulate its growth and development, allowing it to divide and eventually become an entire organism. Since the entire organism was ultimately derived from a single cell with a specific inserted gene, all of the cells in the organism contain that inserted gene.

One of the most common methods of gene editing is CRISPR-Cas9 (CRISPR stands for Clustered Regularly Interspaced Short Palindromic Repeats, and Cas9 stands for CRISPR-associated protein 9). The CRISPR-Cas9 system is extremely popular compared to other gene editing methods because it is comparatively fast, cheap, and accurate. Scientists were inspired to create the CRISPR-Cas9 system by a naturally occurring genome editing system that is used by bacteria. Bacteria infected by viruses capture small pieces of the viruses' DNA and insert them into their own DNA to create segments known as CRISPR arrays, which allow the bacteria to recognize the viruses' DNA in the future and better defend against them. In CRISPR-Cas9, scientists create a sequence of RNA to guide a target sequence of DNA into a cell's DNA. The guide RNA attaches to the Cas9 enzyme so that the Cas9 enzyme can cut the DNA at the targeted location. Then, the cell's own repair machinery adds the inserted DNA to the genome of the cell, completing the modification of the cell, which then divides and forms the basis for the entire organism.

Entirely relevant to any discussion regarding genetically modified organisms is Insulin, a hormone created in the pancreas that regulates the amount of glucose in the blood and is necessary for human survival. Insulin is particularly relevant for people with type 1 diabetes, a chronic condition in which the pancreas makes little or no insulin. Without insulin, a person's body breaks down its own fat and muscle, resulting in a serious short-term condition known as diabetic ketoacidosis. In this condition, a person develops dangerous levels of ketones in their bloodstream and becomes extremely dehydrated. Since type 1 diabetes does not have a cure, patients with the disease have to manually manage the amount of sugar in their blood, often using human insulin produced from GMOs.

GMO insulin mostly comes from bacteria such as Escherichia coli or yeast such as Saccharomyces cerevisiae. GMO insulin is created by inserting human insulin-producing genes into the genome of one of these hosts. The new DNA of the host is known as recombinant DNA. After the host organism multiplies, its offspring produce insulin using the inserted gene, which is then harvested and purified. Lastly, the GMO insulin product is sold by drug makers like Eli Lilly to consumers who need it for survival. Before GMO insulin was available, traditional methods consisted of extracting insulin from the pancreas of animals like pigs or cows. While animal insulin was lifesaving, it did have notable downsides. For example, some patients developed allergic reactions due to the presence of non-human proteins in the insulin. Additionally, batches of animal insulin had various impurities and varying potency/purity. There were also ethical, cultural, and religious concerns. Thus, compared to these methods of obtaining insulin, GMO insulin is effective, efficient, and safe.

Around 8.4 million people in the United States rely on insulin to survive, while over 150 million people around the world depend on insulin to treat their diabetes. These people have historically struggled, as insulin is a rare and expensive commodity that has only recently declined in price. Historically, the price of insulin increased almost annually, far outpacing inflation. Even with legislation such as the Affordable Care Act being put into place, insulin cost as much as $275 per vial in 2017. For most consumers, however, the net price has always been much lower than this figure, as insurance and healthcare cover most of the cost. In 2018, it was estimated that the average net price for one vial of insulin in the United States was $98.70 - while significantly less than $275, this figure was still extremely high and was set with profit in mind, not the livelihoods of American consumers of insulin.

Another reason why insulin has been so expensive in the past is evergreening, the practice of manufacturers slightly changing the formulation of a drug in order to produce a completely brand-new patent on what is just a slightly modified old drug. Studies have indicated that evergreening is hindering the ability of other corporations to compete with drug makers who participate and who hoard patents. Evergreening hurts consumers the most, however, by discouraging generic drugs from being developed - when all of the possible drugs are patented, no organization can step in to provide a generic solution.

For many Americans, the cost of insulin has recently decreased due to the Inflation Reduction Act, which put a cap on the cost of insulin at just $35 for approximately four million seniors with diabetes who are on Medicare. Signed on August 16th, 2022, and touted as the "single largest investment in climate and energy in American history" by the United States Federal Government, the act focused on enabling America to tackle the climate crisis while positioning America to achieve a net-zero economy by 2050 (meaning that America's economy would not contribute negatively to global carbon emissions by 2050). Additionally, the act appropriated billions of dollars to loans for private corporations to achieve climate goals. Importantly, the Inflation Reduction Act provided meaningful financial relief for the millions of seniors on Medicare and strengthened and ensured the successful future of Medicare. Overall, studies indicate that insulin price and affordability were concerns to about 7 in 10 Medicare beneficiaries, meaning the Inflation Reduction Act positively affected the lives of 2.8 million Americans - a third of all Americans who rely on insulin to survive. Estimates indicated that these Americans will save an estimated $734 million dollars annually due to the effects of the Inflation Reduction Act. Furthermore, the Inflation Reduction Act's chief objective was to reduce the federal government budget deficit while curbing inflation, meaning the bill was not only a social positive (helping millions of Americans afford insulin) but was also conscious of the economic effects of increasing government spending in some arenas. However, as is the case with all social policy, there was significant opposition to the bill. Specifically, influential lawmakers and economists claimed that the bill consisted of “reckless spending” and would increase consumer prices and inflation. While the specific effect of this bill on inflation is hard to measure, you would think that a bill titled the “Inflation Reduction Act” would reduce inflation to at least *some* extent - but the reality of the situation is that the act likely did not have any large effect on inflation.

Six months after the approval of the Inflation Reduction Act, an equally significant shift in the landscape of insulin prices occurred, as the three major insulin manufacturers in the United States - Eli Lilly, Sanofi, and Novo Nordisk - all announced plans to cap the price of insulin to just $35 a month, representing a huge price decrease from just five years earlier. The plans to cap and cut the price of insulin went into effect in January of 2024 and have since provided critical relief to the millions of Americans who rely on insulin.

The government’s approval of the Inflation Reduction Act was a major factor that spurred these massive price cuts from private corporations, which have provided life-changing benefits for millions of Americans.

Even before the act was signed, major drug manufacturer Sanofi planned to reduce the cost of their products for uninsured Americans in a preemptive strike due to democratic senators looking to pass the failed "Insulin Act." The actualization of the plans to lower the cost of insulin in the Inflation Reduction Act put more pressure on corporations to lower the cost of insulin for uninsured Americans in the form of public shaming, which played a major role in the decision from Eli Lilly, Sanofi, and Novo Nordisk to cut prices. The public shaming undermined employee morale and negatively affected sales of newer products, hurting the company in the long term more than the huge profit margins from insulin were helping. Further shaming coming from the congressional floor compounded these effects. Particularly, as Eli Lilly was the first company to cut the prices of insulin substantially, shaming from the public and Senator Bernie Sanders convinced Novo Nordisk and Sanofi to fall in line, cutting prices of their insulin products by an average of 70%.

Outside of insulin, the benefits of GMOs are significant and span a variety of areas. In GMO crops, fewer pesticides can be sprayed, soil can be tilled less, viruses are a thing of the past, yields can be higher, produce can have longer shelf lives, and crops can even live longer in a wider variety of environments. Studies have consistently shown GMOs have positive economic and environmental effects; all the while, from a consumer's perspective, GMO foods include added nutrients and lower prices, and from a producer's perspective, GMO foods are easier and cheaper to produce. Genetically engineered food is beneficial for all parties involved.

From the perspective of a farm, GMO crops are advantageous for a number of reasons. In one study of GMO vs non-GMO soybeans, farmers who planted GMO crops experienced slightly higher average yields (51.21 bushels per acre as opposed to 49.26 bushels per acre). However, farmers who planted GMO crops reported paying more for seeds ($26.42 per acre as opposed to $18.89 per acre). By far, the largest difference was in the cost of pesticides, as farmers who used GMO crops reported paying nearly 30% less than farmers who grew non-GMO soybeans and held other cost advantages in all aspects of weed management, including soil tilling. Therefore, producing GMO corn was significantly cheaper for farms than producing non-GMO corn ($9 per acre), while the difference in return was statistically insignificant. All the while, many farms that planted GMO corn indicated that they did so in order to help the environment and deliver a product with fewer pesticides. The study shows that producing consumable GMOs is socially and economically beneficial for farms.

Due to the variety of benefits associated with them, GMO foods are consumed regularly by many Americans. In fact, the majority of Americans, either knowingly or unknowingly, consume foods and food products that come from GMO crops, including oils, sugars, starches, and even fruits and vegetables. GMO crops are so common, in fact, that 92% of all corn planted in 2020 was genetically modified, and GMO products made up above 90% of all soybean, cotton, canola, and sugar beets planted and harvested in 2013 and 2020. Including countries besides the United States, the five most planted GMO crops are soybean (which is used for food for animals, ingredients in processed foods, and making soybean oil), maize (a commonly grown food that is used to feed livestock and humans and is present in many processed foods), cotton (for the textile industry), canola (for cooking oil and margarine), and alfalfa (for feeding cattle). Genetically modified versions of these crops are so popular because the crops themselves are extremely popular throughout a wide variety of global diets while possessing specific traits that allow them to benefit from genetic modification. For example, genetically modified soybeans and canola are herbicide resistant, while genetically modified maize and cotton resist insects. Alfalfa is significantly less commonly planted than soybean, maize, cotton, and canola; it is a crucial crop for livestock feed and may be genetically modified for improved yields and resistance to weeds. The countries with the most land dedicated to planting GM plants are the US, Brazil, Argentina, India, and Canada. These countries produce so many GM plants because they have relaxed government policies that foster biotechnology innovation, they have plenty of land for crops to be planted on, and they have advanced technology that allows them to develop and adopt GMO technologies effectively. The top five countries in terms of GM crop production all rank top eight in the world in terms of size, perhaps indicating that the size of the countries is the reason they produce so many GM plants.

Another extremely popular example of a GMO is pink pineapple, also known as Pinkglow. Pinkglow is a product of food giant Del Monte that is grown in Costa Rica and shipped around the United States (excluding Hawaii, where GM foods are banned). The product is a pink pineapple, which gains its pink color due to an extra dose of lycopene, the same compound that gives watermelon and tomatoes their red color. In regular pineapples, lycopene is converted into beta-carotene by an enzyme. In the pink pineapples, the enzyme is muted, thus causing lycopene to build up and change the color of the fruit. The pineapple was engineered through a technique known as RNA interference, wherein scientists added an antisense RNA that binds to the RNA carrying the message to build the enzyme that converts lycopene to beta-carotene, thus muting it. Lycopene is known to combat oxidative stress and may lower the risk of chronic diseases like heart disease, cancer, and type 2 diabetes; thus, there is an argument to be made that pink pineapples are healthier than their yellow counterparts. Pinkglow is a successful product for Del Monte because it is highly marketable due to its unique and recognizable pink color, its reputation as being healthy, and its taste, which is slightly sweeter and less bitter than regular pineapples. Further rationale for the creation of Pinkglow is that it showed that GM foods could be safe and financially successful, paving the way for other GMO foods to hit the shelves in the future. Another positive impact of Pinkglow is its sustainability - before shipping each Pinkglow product, Del Monte removes the crown of the pineapple because the crowns are used to propagate the fruit. Therefore, Del Monte doesn't need to perform the gene editing process for Pinkglow multiple times, positively contributing to the long-term sustainability of the crop.

GENETICALLY MODIFIED ORGANISMS: THE SOLUTION TO WORLD HUNGER?

GMO food is not simply a promising way for consumers and producers to save money on food, but rather an entirely necessary part of the solution of meeting the future global needs for food. In fact, GMO technology advancement has been argued to be even more important than traditional agricultural advancements and practices in the fight to feed the 9 billion people that will soon inhabit our earth. GMO foods are necessary at a point in which humanity's population grows significantly; however, even with GMO foods being cheap and nutritious compared to non-GMO foods, as many as 783 million people worldwide are still suffering from the devastating effects of hunger. If GMO foods are the answers to the world's future hunger problems, as studies have shown they seem to be, why have they failed to address many systemic problems that the world faces today?

Important to understanding world hunger is understanding that there is not a global food shortage. Indeed, there is more than enough food produced in the world to feed everyone on the planet sufficiently. GMOs provide potentially higher quality food for smaller prices in areas where they are available, but this is useless in the fight against global hunger since global hunger is caused predominantly by poor distribution of food around the world. Certainly, GMOs can take a key step in fighting against world hunger as they can provide a cheaper and more sustainable source of food - but the real issue is distributing these cheaper GMO products to the geographical areas that suffer from world hunger the most. Even if these GMO products were distributed to different geographical areas, they still might not be cheap enough, as most people who suffer from hunger are well below the poverty line. Compounding these effects is the fact that conflict is the biggest driver of hunger, indicating that ongoing global conflicts would likely prevent GMO foods from reaching the right people. While GMOs can help alleviate some of the effects of world hunger by providing a cheap, sustainable alternative to non-GMO food, the answer to ending world hunger lies in geopolitical policies, not genomes.

Further compounding GMOs' ability to fight issues like world hunger is the fact that many people continue to perceive GMOs as unhealthy, unnatural, and unsafe, with some going so far as to avoid GMO products entirely. The average American's view of GMOs tends to be negative, as analysis of social media indicates that the emotions associated with GMOs include "disgust," "sadness," "anger," and "surprise." Further studies show that the average American is willing to pay an average of 30% more for a non-GMO product, despite numerous studies that have shown that GMO products are beneficial for the environment, superior in taste to non-GMO counterparts, and, at worst, neutral for human health. The public's lack of trust in the healthiness of GMO products partially stems from a noted lack of trust in the regulatory processes behind GMOs. The lack of trust from the American public has significantly impeded the progress of GM foods from the lab to the market, as GMO foods are unnecessarily facing difficulties competing with non-GMO foods in the free markets. Improving trust in regulations and labeling is one of the best ways that the government and scientists can improve the public perception of GMOs, which would result in fewer roadblocks for GMO production. Improving the public perceptions of GMOs is an important task for scientists - perhaps as important as the scientific development of GMOs altogether - if GMOs are going to address projected challenges in food production that arise from population growth estimates.

In the United States, GMOs are regulated by the Food & Drug Administration (FDA), the Environmental Protection Agency (EPA), and the Department of Agriculture (USDA). Regulation working through three separate agencies ensures that GMOs are having positive effects on the environment, human health, animal health, and plant health. Overall, the work of these three agencies ensures that GMO foods are as safe as non-GMO foods. Labels that disclose that a food item contains a bioengineered ingredient legally must be present on the package or label of a food, in accordance with regulations set by the FDA. However, food sold by small manufacturers or served by restaurants, food trucks, or other small vendors does not have to disclose its bioengineered status. Importantly, the labels do not contain any information about the safety of the food, simply whether or not it contains bioengineered ingredients. Unfortunately, the FDA's disclosure policy contains numerous loopholes. For example, it excludes highly refined products from mandatory disclosures in most cases. It is estimated that about 70% of GM foods do not actually have to be labeled appropriately due to loopholes. These loopholes likely contribute to the public's lack of trust in the regulation of GMO products. If the loopholes were closed, and all businesses (including restaurants) were forced to disclose their use of GMO products, the American public would trust the government more to effectively regulate GMOs, thus resulting in more Americans considering eating GMOs and realizing the various benefits of GMOs.

In conclusion, GMOs play a vital role in modern agriculture, offering solutions to some of the world's most important challenges. Despite skepticism and regulatory issues, GMOs have been shown to be beneficial and, in the case of GMO insulin, life-saving. While not a cure-all, GMOs remain essential in meeting the demands of a growing population; closing regulatory gaps in order to foster public trust is a crucial next step to maximize GMOs' potential to address global challenges.

Chapter 5

Smoking's Butt End: Chemicals, Carcinogens, and Causes

SMOKING: PROVEN DANGERS

Smoking, whether through traditional cigarettes or novel e-cigarettes and vapes, is pervasive throughout American culture. In the status quo, nearly 30 million Americans regularly smoke cigarettes, either unbothered by or unaware of the dangers they are bringing upon their own health and the health of those who inhale their secondhand smoke. This chapter will delve into smoking's ability to cause cancer and damage various aspects of a person's health while discussing factors that have led many people to smoke on a regular basis and comparing traditional smoking with novel technologies. This chapter will conclude with discussions regarding potential solutions to the smoking epidemic as we seek to put a stop to the leading cause of preventable death in the United States.

A carcinogen is any substance, organism, or other agent that is capable of causing cancer, whether it occurs naturally in the environment or is generated by humans. Most carcinogens work by interacting with a cell's DNA to produce mutations. A few examples of natural carcinogens include ultraviolet rays and viruses such as HPV. A few examples of proven carcinogens created by humans include exhaust fumes and cigarette smoke. Cigarette smoke, the focus of this chapter, is particularly dangerous because it contains an array of diverse chemicals that can cause cancer, including formaldehyde, lead, arsenic, benzene, and polycyclic aromatic hydrocarbons (PAHs). PAHs are harmful to humans in a few different ways, the most significant of which is the toxicity of their reactive metabolites, or small molecules that are intermediate products of metabolism. When a PAH metabolite binds to a person's DNA at a site critical to the regulation of cell differentiation or growth, the DNA might be rendered ineffective. If the DNA is a tumor-suppressor gene, then cancer can form more easily. Other cells most affected by PAH exposure include those in bone marrow, skin tissue, and lung tissue, as these cells have a rapid replicative turnover that can be heavily disturbed by PAH metabolites. Furthermore, certain people may be more genetically susceptible to PAH carcinogens than others as there are certain pre-existing conditions that PAH's carcinogenic ability could worsen. Today, we recognize PAH as one of cigarette smoke's 70 carcinogens that can cause lung cancer.

While it had been difficult for scientists to establish a clear link between smoking and lung cancer in the past, smoking is now widely recognized as a cause of lung cancer. Today, the scientific consensus is that approximately 90% of lung cancers can be attributed to smoking, and general tobacco use is responsible for about 22 percent of all cancer deaths. Furthermore, cigarette smoking is the leading cause of preventable disease, disability, and death in the United States. Smoking increases a person's risk of lung cancer because, as a person inhales tobacco smoke, thousands of chemicals enter their lungs, 70 of which are known carcinogens. While a person's body can repair the damage that's done by these chemicals over time, most smokers participate in smoking for years, accumulating more damage than their body can heal. This often results in the formation of lung cancer cells, as evidenced by the aforementioned statistics. Aside from smoking's strong connection with lung cancer, smoking can also damage a person's lungs in other ways. Specifically, smoking can damage a person's alveoli - tiny air sacs that are the center of the respiratory system's gas exchange. Over time, this damage can lead to chronic obstructive pulmonary disease.

Altogether, the scientific consensus is clear: smoking is extremely damaging to a person's lungs. However, millions of people in America continue to smoke cigarettes each and every day. While there are undoubtedly various social and addictive factors that cause Americans to want to smoke, part of the reason why many Americans continue to partake in the dangerous habit is that, until recently, it had been difficult to establish a truly causal relationship between smoking and cancer, beyond a simple correlation. Luckily, while large tobacco companies shrouded the truth in mystery with deceptive ads targeting America's youth, scientists were able to establish cigarette smoking as the leading cause of lung cancer using four distinct pieces of evidence.

Firstly, a parallel rise in cigarette consumption and lung cancer was noticed by the 1930s, proving that there was a link between lung cancer and cigarette consumption "beyond a reasonable doubt" by 1954. Then, animal experimentation was a key step in proving causality. An influential experiment by Ernst Wynder, Evarts Graham, and Adele Coninger in 1953 showed that artificially painted cigarette smoke caused tumors to develop in mice. This discovery caused tobacco and cigarette companies to fall in value drastically. As the CEOs of these companies planned to refute the growing pile of evidence against their products, a third line of evidence for cigarettes causing lung cancer began to form in the field of cellular pathology. As cellular pathologists explored whether damage from smoking could be discerned at the cellular level, it was determined that precancerous changes could be detected in the cells of smokers - even those who had died from other causes. A final piece of evidence stemmed from the discovery of known carcinogens in cigarette smoke - specifically, PAHs. As summarized in 1961, carcinogens were found in "practically every class of compounds in cigarette smoke." While by 1970, the link between smoking and lung cancer was well-established, misinformation campaigns from tobacco companies, in addition to social stigmas, kept the tobacco market afloat.

Aside from misconceptions regarding the negative health effects of smoking, a huge reason why many people continue to smoke is the addictive nature of cigarettes and other smoking devices. Products containing tobacco are addictive due to nicotine, a chemical compound that speeds up the messages that travel between the brain and the body. Nicotine gives its user mild stimulation, relaxation, and increased ability to concentrate. However, it can also result in bad breath, dizziness, confusion, headaches, and even seizures. In the long term, nicotine results in almost exclusively negative effects, heightening a user's risk of stroke, blindness, infertility, and addiction. Thus, the negative consequences of smoking exist not only in its carcinogens - the addictive nature of tobacco itself, due to the chemical compound nicotine, causes negative effects to those who choose to smoke it. At the same time, however, nicotine is significantly less dangerous and carcinogenic than many of the other chemicals inhaled by someone who smokes tobacco. Therefore, some people have resorted to nicotine replacement therapies, such as skin patches, gum, and lozenges, which can give people who are trying to quit smoking a small amount of nicotine, reducing their yearning to smoke. Since nicotine may be as addictive as heroin and cocaine, drastic measures are often necessary in order to get people off of the dangerous substance.

When a person smokes a cigarette, they are taking in approximately 1-2 milligrams of nicotine, which quickly moves into their bloodstream. The nicotine stimulates the person's adrenal glands, resulting in the discharge of epinephrine, also known as adrenaline. This rush of adrenaline released by nicotine stimulates the body, resulting in many of the other effects of nicotine: an increase in blood pressure, focus, respiration, and heart rate. This element of nicotine is not unique; most drugs contain compounds that can spike a person's adrenaline. What makes nicotine particularly dangerous is its effects on the brain: when a person ingests nicotine, their brain will release feelings of pleasure due to the compound's ability to change the brain's amount of neurotransmitter dopamine, which is usually broken down by neurons in the prefrontal cortex. Nicotine also targets the brain's pain relievers and can block the action of endogenous opioid peptides in the brain, causing reduced perceptions of pain and stress. Overall, nicotine hard-wires the brain to crave it while causing pleasurable feelings of relaxation while releasing adrenaline, causing smokers to develop an addiction to it, which can be difficult to stop.

As shown, smoking is extremely dangerous due to its carcinogenic effects, its addictiveness, and other dangers associated with smoking, such as headaches and negative effects on lung health. With recent advances in technology, electronic cigarettes and vapes have introduced a new means by which a person can smoke, involving a battery-powered vape device that creates an aerosol that looks like water vapor but actually contains nicotine, flavoring, and more than 30 other chemicals. While vaping is often considered to be safer than traditional smoking, this is heavily debated as vaping still contains its own carcinogens and dangerous chemicals, such as acetone. Additionally, vaping is extremely addictive as a single vape pod could contain as much nicotine as twenty cigarettes. Therefore, although many people think of vapes as a positive alternative to smoking, vaping actually has overall negative effects as it attracts people whom smoking might not attract while being even more addictive and, at best, slightly less dangerous. Studies have shown that vaping makes teens more likely to start smoking and become addicted to nicotine while failing to help people quit smoking when compared to traditional means of quitting smoking, such as using nicotine patches. Additionally, vapes can act as a gateway drug; since some vapes can contain cannabis, vaping can make people who smoke more open to the idea of trying cannabis in the future. It is well understood that cannabis acts as a gateway drug to deadly drugs such as heroin and methamphetamine, meaning that vaping, while seemingly innocuous and innocent when compared to smoking traditional cigarettes or smoking weed, can actually make people more likely to endanger their lives with even more addictive drugs.

Unfortunately, a person can become a victim of the dangers of smoking without ever picking up a cigarette themselves. "Secondhand smoke" is smoke that comes from burning tobacco products such as cigarettes. Even brief exposure to secondhand smoke can give a person exposure to dangerous chemicals such as benzene and ammonia and can open a person up to the same dangers and carcinogenic factors that a person who does smoke is exposed to. Secondhand smokers are at a higher risk of heart disease, stroke, lung cancer, and premature death compared to those who live in smoke-free areas. Additionally, the harms of secondhand smoke are immediate - secondhand smoke exposure produces harmful inflammatory and respiratory effects within 60 minutes of exposure. Furthermore, it is estimated that since 1964, about 2.5 million people who did not themselves smoke died from problems caused by secondhand smoke exposure. The dangers of second-hand smoking from cigarettes are well-defined and comparable to the dangers of smoking directly; however, the dangers of secondhand e-cigarettes are less clear. While research indicates that e-cigarettes emit some harmful substances in aerosols containing nicotine, formaldehyde, and metals, more research is needed to fully understand the negative impact of emissions from e-cigarettes on human health. As e-cigarettes and vapes do not produce the same thick clouds of smoke that smoking does, there are certainly differences in risks that should continue to be evaluated. While it seems like e-cigarettes may be comparatively safer to consume in the form of secondhand smoke, the aerosol emitted from vapes and e-cigarettes is still dangerous and carcinogenic. Unfortunately, there is little that an individual person can do in order to avoid second-hand smoke, as, counterintuitively, rural Americans are even more likely to be exposed to disease or disability due to secondhand smoke than urban Americans because they lack access to healthcare. It seems as though aggressive government and social policy will be the only way to clean our air of carcinogenic and dangerous smoke.

SMOKING: POTENTIAL SOLUTIONS

As previously explained, smoking is extremely addictive, whether it comes in the form of cigarettes or vapes, and can be extremely difficult to quit. Therefore, the best way to prevent the cancers and deaths that smoking, vaping, and secondhand smoke inhalation cause is to ensure that people never start smoking or vaping in the first place. One way that the government can prevent people from smoking or vaping in the first place is by establishing aggressive sin taxes. Sin taxes are simply taxes levied on certain goods at the time of purchase. However, sin taxes can uniquely apply to items that are deemed morally suspect, harmful, or costly to society. Sin taxes provide revenue to the government while de-incentivizing harmful behaviors. Simply put, if cigarettes or vaping devices are made more expensive due to a heavy sin tax, people would be less likely to buy them in the first place. Sin taxes have already been proposed in various places and on various goods and services, including alcohol, gambling, cigarettes, vaping, and pornography. While sin taxes may deter some people from smoking or vaping, they are not a perfect solution. Firstly, they unfairly target people who are impoverished while not making a big impact on those who are wealthier. While smoking and vaping are dangerous, a sin tax might make them become viewed as a luxury that can only be afforded by the wealthy. Another problem with sin taxes is if someone is already addicted to nicotine, they may find themselves in a position where they are giving up a lot of their money or personal belongings in order to afford to buy more cigarettes or vaping devices.

Another method that could be used to prevent smoking is social reform. Even though many people smoke because they are addicted or they feel like it helps give them stress relief or pleasure, most people begin smoking because of social pressures; they believe that smoking "looks cool" or will help them fit in. Many kids begin smoking because they think that smoking is a way for them to show their independence and rebelliousness while fitting in with their peers who also smoke. This is largely because of clever and deceptive marketing tactics from the tobacco industry, which specifically target teenagers to make them think that they should smoke. Nationwide ad campaigns that discuss the dangers of smoking could make children and teenagers think that smoking is more negative. Additionally, more education in school about the dangers of smoking would help kids understand that smoking comes at a cost. Finally, the government could ban ways of marketing smoking and tobacco that directly appeal to children.

A final method that could be used to prevent smoking is outright banning. While outright bans would reduce the amount of cigarettes and vaping devices on the market, it is likely that this method of preventing smoking would not be successful for a few reasons. Firstly, it is likely that an underground black market would form surrounding the sale of cigarettes and vaping devices. In this black market, since there would be no government regulation, dangerous substances such as fentanyl could find their way into cigarettes or vapes. This would make smoking even more dangerous for those addicted, and those addicted would not be able to stop even with the increased dangers in mind because of the addictive power of nicotine. Another reason why an outright ban on cigarettes and vapes would not be successful is that this ban would not be very popular; in the United States, a proposal to ban menthol-flavored cigarettes was met with an immense amount of negative feedback and a historic level of attention due to people who are addicted to smoking and those who believe that a ban of cigarettes and vapes would be authoritarian and un-American.

In conclusion, smoking remains a pervasive issue in American society, contributing significantly to lung cancer, preventable death, and health complications. Smoking's carcinogenic effect, coupled with nicotine's addictive effect, creates a dire situation with an urgent need for some form of intervention. Strategies such as sin taxes, social reforms, educational reforms, and government bans are all promising ways in which we can work to curb smoking rates; while none of these methods are perfect solutions, concerted efforts at the individual, societal, and governmental levels are essential to combating the smoking epidemic and fostering healthier and cleaner communities for generations to come.

Chapter 6

What is "Global Health," Anyways?

With a title emphasizing *global health*, an astute critic might point out that the solutions discussed within this book are all centered to the United States; although issues like HIV and AIDS are most problematic and stigmatized in countries like China and Southern Africa, this book primarily considers the social, economic, and scientific context specific to the United States. Why, then, does a book that claims to be about global health, in a conversation detailing global issues, talk almost exclusively about the United States?

In third-world countries to which the United States supplies "aid", the aid being sent (particularly in the context of health and medicine) is often machines and other utilities that were too old or out-of-fashion to be used in the United States, and instead of being thrown out, these utilities wind up being "donated" to other less-wealthy countries. This is the old way of "doing" global health - applying solutions that might work in one country to another. Perhaps, in an under-funded hospital in America, those machines and utilities might have been useful. The assumption is made that other countries have the same need as our own country and, thus, we can send them these utilities to improve their facilities and save lives. The reality of the situation, however, is that the machines that are sent to other countries as aid most often wind up completely wasted, collecting dust in a closet filled with tools that either don't fit in with a facility's existing infrastructure or don't work whatsoever.

This situation is analogous to providing any sort of aid to other countries - without looking at specifically what the situation is in the other countries, it is difficult to provide effective solutions. While, in theory, it would be easy to advocate for global social reform in order to limit the amount of smoking worldwide, the specific social reform necessary differs greatly by country. In the United States, as explained, social reform by means of stopping advertising to children is necessary. In other countries, for example, smoking might be an issue for different reasons. In China, cigarettes are often used as gifts and illustrate someone's respect and hospitality. While this is different than the situation in the United States, there are some similarities - for example, like in the United States, teenagers in China see smoking as a sign of maturity. In order for someone to advocate for social reform in China, they must be extremely familiar with the social implications of implementing public policy, in addition to considering economic and environmental factors.

Thus, creating recommendations for world-wide public policy is difficult, if not impossible. Perhaps this, in part, is one of the difficulties with the Sustainable Development Goals. The goals are truly attempting to better society for both the people and the planet - but they do so in such a way that is unactionable on a global scale. This isn't a fault of any of the specific goals themselves, but instead a fault of what the goals are trying to accomplish. While the most ambitious thing about the goals seems, upon first glance, to be the depth of the specific objective in interdisciplinary arenas such as climate change, income inequality, and environmental justice, the true challenge that the goals face is being implemented in a global scale.

In order to truly "end hunger" (goal two), an organization would have to go to every single country, perform qualitative and quantitative analyses in order to determine the root cause of hunger, and then enact substantial policy on a country-by-country (or even state-by-state) basis while considering social and economic factors. Despite being one of the most worthwhile challenges that humankind faces, it is not possible for a single organization to end hunger on a global scale with our current global socioeconomic infrastructure.

Surprisingly, this is not because the resources to end hunger do not exist: as explained in Chapter 4, the food exists, and is grown, and is distributed poorly, and is wasted. In order to truly solve world hunger, and the other issues outlined by the sustainability goals, collaboration is necessary - and not collaboration in the sense of the "old way of doing global health" - collaboration in the sense of a new, transformative approach involving coordinated efforts across governments, NGOs, private enterprises, and local communities. Additionally, progress is necessary: not just progress in the realm of science and social science, but multidisciplinary and equitable progress involving technological innovation, economic development, cultural understanding, and environmental sustainability.

Addressing global health challenges necessitates a paradigm shift from outdated, one-size-fits-all solutions to a nuanced, collaborative approach that respects and integrates the unique social, economic, and cultural contexts of each country.

Bibliography

Antibiotic Resistance and its "Solutions"

2019 Antibiotic Resistance Threats Report. (2019, December). Centers for Disease Control and Prevention. https://www.cdc.gov/drugresistance/biggest-threats.html

35,000 annual deaths from antimicrobial resistance in the EU/EEA. (2022, November 17). European Centre for Disease Prevention and Control. https://www.ecdc.europa.eu/en/news-events/eaad-2022-launch

Sifferlin, A. (2015, November 16). *5 common myths about antibiotic-resistant bacteria*. Time. https://time.com/4114345/antibiotic-resistance-common-myths/

An estimated 1.2 million people died in 2019 from antibiotic-resistant bacterial infections (2022, January 20). University of Oxford. https://www.ox.ac.uk/news/2022-01-20-estimated-12-million-people-died-2019-antibiotic-resistant-bacterial-infections

Antibiotic resistance. (n.d.). Cedars-Sinai. Retrieved October 10, 2023, from https://www.cedars-sinai.org/health-library/diseases-and-conditions/a/antibiotic-resistance.html

Antibiotic resistance. (2020, July 31). World Health Organization. https://www.who.int/news-room/fact-sheets/detail/antibiotic-resistance

Antibiotic resistance. (2023, March). National Foundation for Infectious Diseases. Retrieved October 10, 2023, from https://www.nfid.org/antibiotic-resistance/

Antibiotics. (n.d.-a). MedlinePlus. Retrieved October 10, 2023, from https://medlineplus.gov/antibiotics.html

Antibiotics: Are you misusing them? (2023, July 11). Mayo Clinic. https://www.mayoclinic.org/healthy-lifestyle/consumer-health/in-depth/antibiotics/art-20045720

Be Antibiotics Aware. (2021, November 12). Centers for Disease Control and Prevention. https://www.cdc.gov/patientsafety/features/be-antibiotics-aware.html

Beceiro, A., Tomás, M., & Bou, G. (2013). Antimicrobial resistance and virulence: a successful or deleterious association in the bacterial world?. *Clinical microbiology reviews*, *26*(2), 185–230. https://doi.org/10.1128/CMR.00059-12

Britannica, The Editors of Encyclopaedia (2023, October 13). *natural selection. Encyclopedia Britannica*. https://www.britannica.com/science/natural-selection

Duncan, C. (2022, February 24). Scientists discover new approach to fighting antibiotic resistance. *Imperial News; Imperial College London*. https://www.imperial.ac.uk/news/234060/scientists-discover-approach-fighting-antibiotic-resistance/

Felman, A., & Begum, F. (2019, January 18). What to know about antibiotics. *MedicalNewsToday*. https://www.medicalnewstoday.com/articles/10278

Floersh, H. (2023, June 2). Scientists use AI to find new antibiotics, but can they climb out of the "valley of death" to the clinic? *Fierce Biotech*. https://www.fiercebiotech.com/research/ai-finds-new-drug-curbs-acinetobacter-mice

Goode, J. (2023, March 29). As superbug infections grow more common, the world is running out of drugs to treat them. *NBC News*. https://www.nbcnews.com/health/health-news/-arrived-po

st-antibiotic-era-warns-new-drugs-deadly-superbugs-rcna76601

Goodman, B. (2023, May 25). A new type of antibiotic, discovered with artificial intelligence, may defeat a dangerous superbug. *CNN*. https://www.cnn.com/2023/05/25/health/antibiotic-artificial-intelligence-superbug/index.html

Gould, I. M., & Bal, A. M. (2013). New antibiotic agents in the pipeline and how they can help overcome microbial resistance. *Virulence*, *4*(2), 185–191. https://doi.org/10.4161/viru.22507

How Antimicrobial Resistance Happens. (2022c, October 24). Centers for Disease Control and Prevention. https://www.cdc.gov/drugresistance/about/how-resistance-happens.html

Kadner, R. J. and Rogers, . Kara (2023, October 1). *bacteria*. *Encyclopedia Britannica*. https://www.britannica.com/science/bacteria

Lack of innovation set to undermine antibiotic performance and health gains. (2022, June 22). World Health Organization. https://www.who.int/news/item/22-06-2022-22-06-2022-lack-of-innovation-set-to-undermine-antibiotic-performance-and-health-gains

Levin, B. R. (1996). The Evolution and Maintenance of Virulence in Microparasites. *Emerging Infectious Diseases*, *2*(2), 93-102. https://doi.org/10.3201/eid0202.960203.

LiverTox: Clinical and Research Information on Drug-Induced Liver Injury [Internet]. Bethesda (MD): National Institute of Diabetes and Digestive and Kidney Diseases; 2012-. Penicillins (1st Generation) [Updated 2020 Oct 20]. Available from https://www.ncbi.nlm.nih.gov/books/NBK548801/

Loree J, Lappin SL. (2023, August 14). Bacteriostatic Antibiotics. *StatPearls Publishing*. Retrieved October 10, 2023, from https://www.ncbi.nlm.nih.gov/books/NBK547678/

MacLean, C. (2019, September 13). How can evolutionary biology help to get rid of antibiotic resistant bacteria? *University of Oxford*. Retrieved October 10, 2023, from https://www.ox.ac.uk/news/science-blog/how-can-evolutionary-biology-help-get-rid-antibiotic-resistant-bacteria

Metz, M., & Shlaes, D. M. (2014). Eight more ways to deal with antibiotic resistance. *Antimicrobial Agents and Chemotherapy*, *58*(8), 4253–4256. https://doi.org/10.1128/AAC.02623-14

Mullard, A. (2023). FDA approves new antibiotic combination for drug-resistant pneumonia. *Nature Reviews Drug Discovery, 22*(7), 527–527. https://doi.org/10.1038/d41573-023-00096-8

Mutations and selection – Antibiotic resistance. (n.d.). ReAct. Retrieved October 10, 2023, from https://www.reactgroup.org/toolbox/understand/antibiotic-resistance/mutation-and-selection/

National Infection & Death Estimates for Antimicrobial Resistance. (2022a, July 6). Centers for Disease Control and Prevention. https://www.cdc.gov/drugresistance/national-estimates.html

New antibiotic from microbial "dark matter" could be powerful weapon against superbugs. (2023, August 22). ScienceDaily. Retrieved October 10, 2023, from https://www.sciencedaily.com/releases/2023/08/230822111734.htm

New antibiotic, clovibactin, kills bacteria without developing resistance. (2023, August 22). Genetic Engineering and

Biotechnology News. https://www.genengnews.com/topics/infectious-diseases/new-antibiotic-clovibactin-kills-bacteria-without-developing-resistance/

Parkinson, J. (2023, May 24). *What a Newly FDA Approved Antibiotic Means to Clinicians, Patients*. ContagionLive. https://www.contagionlive.com/view/what-a-newly-fda-approved-antibiotic-means-to-clinicians-patients

Pion, S. (n.d.). Antibiotics: Know when you need them. *Atrium Health*. Retrieved October 10, 2023, from https://atriumhealth.org/medical-services/prevention-wellness/antibiotics

Seaton, J. (2023, July 11). AI could quickly screen thousands of antibiotics to tackle superbugs. *Scientific American*. Retrieved October 10, 2023, from https://www.scientificamerican.com/article/ai-could-quickly-screen-thousands-of-antibiotics-to-tackle-superbugs/

Talkington, K. (2019, November 13). How to combat antibiotic resistance: 5 priorities for 2020. *The Pew Charitable Trusts*. https://pew.org/2CDi3WI

Trafton, A. (2020, February 20). Artificial intelligence yields new antibiotic. *MIT News; Massachusetts Institute of Technology*. https://news.mit.edu/2020/artificial-intelligence-identifies-new-antibiotic-0220

Wagner, R. R. and Krug, . Robert M. (2023, September 14). *virus*. *Encyclopedia Britannica*. https://www.britannica.com/science/virus

About Antimicrobial Resistance. (2022b, October 5). Centers for Disease Control and Prevention. https://www.cdc.gov/drugresistance/about.html

What is antibiotic resistance. (n.d.). Missouri Department of Health. Retrieved October 10, 2023, from https://health.mo.gov/safety/antibioticresistance/generalinfo.php

WHO publishes list of bacteria for which new antibiotics are urgently needed. (2017, February 17). World Health Organization. https://www.who.int/news/item/27-02-2017-who-publishes-list-of-bacteria-for-which-new-antibiotics-are-urgently-needed

The Road to Immunity: A Comprehensive Exploration of HIV

About HIV. (2022b, June 30). Centers for Disease Control and Prevention. https://www.cdc.gov/hiv/basics/whatishiv.html

Antibody. (2023b, November 18). National Human Genome Research Institute. https://www.genome.gov/genetics-glossary/Antibody

Coulson, M. (2022, November 29). *Why don't we have an HIV vaccine?* Johns Hopkins - Bloomberg School of Public Health. https://publichealth.jhu.edu/2022/why-dont-we-have-an-hiv-vaccine

Greely H. T. (2019). CRISPR'd babies: human germline genome editing in the 'He Jiankui affair'. Journal of law and the biosciences, 6(1), 111–183. https://doi.org/10.1093/jlb/lsz010

History of AIDS. (2021, February 21). HISTORY. https://www.history.com/topics/1980s/history-of-aids

History of HIV/AIDS. (n.d.). Canadian Foundation for AIDS Research. Retrieved November 19, 2023, from https://canfar.com/awareness/about-hiv-aids/history-of-hiv-aids/

HIV-AIDS: A virus master of evasion. (2023, May 24). Institut Pasteur. https://www.pasteur.fr/en/research-journal/news/hiv-aids-virus-master-evasion

HIV Prevention. (2021, June 1). Centers for Disease Control and Prevention. https://www.cdc.gov/hiv/basics/prevention.html

HIV Testing. (2022a, June 9). Centers for Disease Control and Prevention. https://www.cdc.gov/hiv/testing/index.html

How HIV Infects a Cell. (n.d.). International Partnership for Microbicides. Retrieved November 19, 2023, from https://www.ipmglobal.org/how-hiv-infects-cell

Li, J. R., Walker, S., Nie, J. B., & Zhang, X. Q. (2019). Experiments that led to the first gene-edited babies: the ethical failings and the urgent need for better governance. Journal of Zhejiang University. Science. B, 20(1), 32–38. https://doi.org/10.1631/jzus.B1800624

Murthy G. (2008). The socioeconomic impact of human immunodeficiency virus / acquired immune deficiency syndrome in India and its relevance to eye care. *Indian journal of ophthalmology*, *56*(5), 395–397. https://doi.org/10.4103/0301-4738.42416

Pebody, R. (2021, May 19). *Types of antiretroviral medications*. Aidsmap. https://www.aidsmap.com/about-hiv/types-antiretroviral-medications

Raposo V. L. (2019). *The First Chinese Edited Babies: A Leap of Faith in Science. JBRA assisted reproduction, 23(3), 197–199. https://doi.org/10.5935/1518-0557.20190042*

Refsland EW, Hultquist JF, Luengas EM, et al. Natural Polymorphisms in Human APOBEC3H and HIV-1 Vif Combine to Affect Viral Infectivity and G-to-A Mutation Levels in Primary T Lymphocytes. PLOS Genetics. 2014.

Retrovirus. (2023a, November 18). National Human Genome Research Institute. https://www.genome.gov/genetics-glossary/Retrovirus

Rodríguez-Mora, S., De Wit, F., García-Perez, J., Bermejo, M., López-Huertas, M. R., Mateos, E., Martí, P., Rocha, S., Vigón,

L., Christ, F., Debyser, Z., Vílchez, J. J., Coiras, M., & Alcamí, J. (2019). *The mutation of Transportin 3 gene that causes limb girdle muscular dystrophy 1F induces protection against HIV-1 infection. PLoS pathogens, 15(8), e1007958.* https://doi.org/10.1371/journal.ppat.1007958

The He Lab. (2018, November 25). *About Lulu and Nana: Twin Girls Born Healthy After Gene Surgery As Single-Cell Embryos [Video]. Youtube. https://www.youtube.com/watch?v=th0vnOmFltc*

Ways HIV Can Be Transmitted. (2022c, March 4). Centers for Disease Control and Prevention. https://www.cdc.gov/hiv/basics/hiv-transmission/ways-people-get-hiv.html

What are HIV and AIDS?. (2023, January 13). HIV.gov. https://www.hiv.gov/hiv-basics/overview/about-hiv-and-aids/what-are-hiv-and-aids/

World AIDS Day 2023. (2023). UNAIDS. https://www.unaids.org/sites/default/files/media_asset/UNAIDS_FactSheet_en.pdf

Insulin and Beyond: Exploring Genetically Modified Organisms

A global food crisis. (2023). World Food Programme. https://www.wfp.org/global-hunger-crisis

Agricultural Marketing Service. (n.d.). Information for consumers. Retrieved February 26, 2024, from https://www.ams.usda.gov/rules-regulations/be/consumers

Baeshen, N. A., Baeshen, M. N., Sheikh, A., Bora, R. S., Ahmed, M. M. M., Ramadan, H. A. I., Saini, K. S., & Redwan, E. M. (2014). Cell factories for insulin production. Microbial Cell Factories, 13(1), 141. https://doi.org/10.1186/s12934-014-0141-0

Buse, J. B., Davies, M. J., Frier, B. M., & Philis-Tsimikas, A. (2021). 100 years on: The impact of the discovery of insulin on clinical outcomes. BMJ Open Diabetes Research & Care, 9(1), e002373. https://doi.org/10.1136/bmjdrc-2021-002373

Coffey, D. (2023, September 4). Genetically engineered pink pineapples are flying off shelves: What gives them their distinctive color? Livescience.Com. https://www.livescience.com/planet-earth/plants/pink-pineapples-are-in-high-demand

Collier, R. (2013). Drug patents: The evergreening problem. CMAJ : Canadian Medical Association Journal, 185(9), E385–E386. https://doi.org/10.1503/cmaj.109-4466

Dafny, L. S. (2023). Falling insulin prices—What just happened? New England Journal of Medicine, 388(18), 1636–1639. https://doi.org/10.1056/NEJMp2303279

Duffy, M., & Ernst, M. (2005, October 11). *Does planting GMO seed boost farmers' profits?* https://www.legis.iowa.gov/docs/publications/SD/4416.pdf

Fabricant, F. (2022, February 14). Del Monte pineapples go pink. The New York Times. https://www.nytimes.com/2022/02/14/dining/del-monte-pinkglow-pineapples.html

Inflation Reduction Act of 2022. (2023, September 22). Energy.Gov. https://www.energy.gov/lpo/inflation-reduction-act-2022

Kansteiner, F. (2023, April 20). *What spurred Lilly, Novo and Sanofi to slash insulin prices?* Retrieved February 26, 2024, from https://www.fiercepharma.com/pharma/impetus-behind-lilly-novo-and-sanofis-insulin-price-cuts-explained-report

Kennedy, M. (2022, October 24). Pros and cons of GMOs: An evidence-based comparison of genetically modified foods. Business Insider. https://www.businessinsider.com/guides/health/diet-nutrition/gmo-pros-and-cons

Largest countries in the world by area. (n.d.). Worldometer. Retrieved February 26, 2024, from https://www.worldometers.info/geography/largest-countries-in-the-world/

Legal Appeal Challenges Hidden GMO Foods in Marketplace. (2023, September 6). Center for Food Safety. https://www.centerforfoodsafety.org/press-releases/6850/legal-appeal-challenges-hidden-gmo-foods-in-marketplace

Lovelace, B. Jr. (2023, March 1). Drugmaker Eli Lilly caps the cost of insulin at $35 a month, bringing relief for millions. NBC News. https://www.nbcnews.com/health/health-news/eli-lilly-caps-cost-insulin-35-month-rcna72713

Mahon, S. H. (2023, April 18). Pink Pineapples Are As Magical As They Sound. Southern Living.

https://www.southernliving.com/what-is-a-pink-pineapple-7482353

MedlinePlus. (2022, March 22). What are genome editing and CRISPR-Cas9? https://medlineplus.gov/genetics/understanding/genomicresearch/genomeediting/

Meilan, R. (2016, September 12). What are GMOs? Purdue University - College of Agriculture. https://ag.purdue.edu/gmos/what-are-gmos.html

Muir, W. M. (2017, May 9). What is gene editing? Purdue University - College of Agriculture. https://ag.purdue.edu/gmos/gene-editing.html

National Geographic Society. (2024, January 3). Are genetically modified crops the answer to world hunger? https://education.nationalgeographic.org/resource/are-genetically-modified-crops-answer-world-hunger

Oliver, M. J. (2014). Why We Need GMO Crops In Agriculture. Missouri Medicine, 111(6), 492–507. https://www.ncbi.nlm.nih.gov/pmc/articles/PMC6173531/

Pew Research Center. (2016, December 1). 3. Public opinion about genetically modified foods and trust in scientists connected with these foods. Pew Research Center Science & Society. https://www.pewresearch.org/science/2016/12/01/public-opinion-about-genetically-modified-foods-and-trust-in-scientists-connected-with-these-foods/

Popli, N. (2023, March 2). How the other insulin makers are responding to eli lilly's price cap. TIME. https://time.com/6259974/insulin-eli-lilly-cost-cap-sanofi-novo-nordisk/

Riviera Produce. (2021, August 5). Pinkglow Pineapple. https://www.rivieraproduce.com/pinkglow-pineapple-every

thing-you-need-to-know-about-this-new-summer-sensation/

Sohi, M., Pitesky, M., & Gendreau, J. (n.d.). Analyzing public sentiment toward GMOs via social media between 2019-2021. GM Crops & Food, 14(1), 1–9. https://doi.org/10.1080/21645698.2023.2190294

The history of a wonderful thing we call insulin. (2019, July 1). American Diabetes Association. https://diabetes.org/blog/history-wonderful-thing-we-call-insulin

The White House. (2023, March 2). Fact sheet: President biden's cap on the cost of insulin could benefit millions of americans in all 50 states. The White House. https://www.whitehouse.gov/briefing-room/statements-releases/2023/03/02/fact-sheet-president-bidens-cap-on-the-cost-of-insulin-could-benefit-millions-of-americans-in-all-50-states

Tseng, C.-W., Masuda, C., Chen, R., & Hartung, D. M. (2020). Impact of higher insulin prices on out-of-pocket costs in medicare part d. Diabetes Care, 43(4), e50–e51. https://doi.org/10.2337/dc19-1294

Type 1 diabetes symptoms and treatments. (2023, November 17). NHS Inform. https://www.nhsinform.scot/illnesses-and-conditions/diabetes/type-1-diabetes/

United States Food & Drug Administration. (2022a). GMO Crops, Animal Food, and Beyond. FDA. https://www.fda.gov/food/agricultural-biotechnology/gmo-crops-animal-food-and-beyond

United States Food & Drug Administration. (2022b). Why Do Farmers in the U.S. Grow GMO Crops? FDA. https://www.fda.gov/food/agricultural-biotechnology/why-do-farmers-us-grow-gmo-crops

United States Food & Drug Administration. (2023). How GMOs Are Regulated in the United States. FDA. https://www.fda.gov/food/agricultural-biotechnology/how-gmos-are-regulated-united-states

University of Florida. (2017). *Diabetes: The hidden epidemic*. Retrieved February 26, 2024, from https://healthstreet.program.ufl.edu/wordpress/files/2018/12/The-cost-of-Insulin.pdf

U.S. Department of Health and Human Services. (2023, January 24). New HHS Report Finds Major Savings for Americans Who Use Insulin Thanks to President Biden's Inflation Reduction Act. https://www.hhs.gov/about/news/2023/01/24/new-hhs-report-finds-major-savings-americans-who-use-insulin-thanks-president-bidens-inflation-reduction-act.html

U.S. Department of Health and Human Services. (2024). Inflation reduction act of 2022. https://www.hhs.gov/inflation-reduction-act/index.html

What GM crops are being grown and where? . (2016, May). Royal Society. https://royalsociety.org/news-resources/projects/gm-plants/what-gm-crops-are-currently-being-grown-and-where/

What is diabetes? (2023, April). National Institute of Diabetes and Digestive and Kidney Diseases. https://www.niddk.nih.gov/health-information/diabetes/overview/what-is-diabetes

World hunger facts & statistics. (n.d.). Action Against Hunger. Retrieved February 26, 2024, from https://www.actionagainsthunger.org/the-hunger-crisis/world-hunger-facts/

Smoking's Butt End: Chemicals, Carcinogens, and Causes

Agency for Toxic Substances and Disease Registry. (2023, May 25). Polycyclic Aromatic Hydrocarbons (PAHs). https://www.atsdr.cdc.gov/csem/polycyclic-aromatic-hydrocarbons/pathogenic_changes.html

Alcohol and Drug Foundation. (2024, May). Nicotine. https://adf.org.au/drug-facts/nicotine/

American Cancer Society. (2020, October 28). Harmful Chemicals in Tobacco Products. https://www.cancer.org/cancer/risk-prevention/tobacco/carcinogens-found-in-tobacco-products.html

American Cancer Society. (2023, March 21). Viruses That Can Lead to Cancer. https://www.cancer.org/cancer/risk-prevention/infections/infections-that-can-lead-to-cancer/viruses.html

American Lung Association. (2023, May 31). Why Kids Start Smoking. https://www.lung.org/quit-smoking/helping-teens-quit/why-kids-start-smoking

American Psychological Association. (2016). Smoking and Tobacco Use in Rural Populations. https://www.apa.org/pi/health-equity/resources/smoking-rural-populations

Blaha, M. J. (2022, January 20). 5 Vaping Facts You Need to Know. Johns Hopkins Medicine. https://www.hopkinsmedicine.org/health/wellness-and-prevention/5-truths-you-need-to-know-about-vaping

Centers for Disease Control and Prevention. (2022, September 14). General Information About Secondhand Smoke. https://www.cdc.gov/tobacco/secondhand-smoke/about.html

Centers for Disease Control and Prevention. (2023a, July 31). What Are the Risk Factors for Lung Cancer? https://www.cdc.gov/cancer/lung/basic_info/risk_factors.htm

Centers for Disease Control and Prevention. (2023b, November 2). Fast Facts and Fact Sheets—Smoking and Cigarettes. https://www.cdc.gov/tobacco/data_statistics/fact_sheets/fast_facts/index.htm

Kagan, J. (2023, September). Sin Tax Definition and How It Works. Investopedia. https://www.investopedia.com/terms/s/sin_tax.asp

National Human Genome Research Institute. (2024, May 12). Carcinogen. https://www.genome.gov/genetics-glossary/Carcinogen

NIDA. 2021, April 12. How does tobacco deliver its effects?. Retrieved from https://nida.nih.gov/publications/research-reports/tobacco-nicotine-e-cigarettes/how-does-tobacco-deliver-its-effects on 2024, May 12

NIDA. 2023, October 5. Is marijuana a gateway drug?. Retrieved from https://nida.nih.gov/publications/research-reports/marijuana/marijuana-gateway-drug on 2024, May 12

Proctor, R. N. 2012. The history of the discovery of the cigarette-lung cancer link: evidentiary traditions, corporate denial, global toll. Tobacco Control 21: 87-91

Texas Health and Human Services. (n.d.). What is Vaping? Retrieved May 12, 2024, from https://www.dshs.texas.gov/vaping/what-is-vaping

Tin, A. (2024, April 26). Menthol cigarette ban delayed due to “immense” feedback, Biden administration says. CBS News.

https://www.cbsnews.com/news/menthol-cigarette-ban-delay-fda-feedback-biden-administration/

United States Environmental Protection Agency. (2024, March 2). Secondhand Electronic-Cigarette Aerosol and Indoor Air Quality. https://www.epa.gov/indoor-air-quality-iaq/secondhand-electronic-cigarette-aerosol-and-indoor-air-quality

U.S. Food & Drug Administration. (n.d.). Why Are Tobacco Products So Hard To Quit? https://www.fda.gov/media/157459/download?attachment

U.S. Food & Drug Administration. (2022, June). Nicotine Is Why Tobacco Products Are Addictive. https://www.fda.gov/tobacco-products/health-effects-tobacco-use/nicotine-why-tobacco-products-are-addictive

Yale Medicine Magazine. (2001). How Nicotine May Buffer the Brain. Yale School of Medicine. https://medicine.yale.edu/news/yale-medicine-magazine/article/how-nicotine-may-buffer-the-brain/

Yetman, D. (2021, March 29). The Connection Between Smoking and Lung Cancer. Healthline. https://www.healthline.com/health/lung-cancer/smoking-lung-cancer

https://www.[illegible] [illegible]

United States [illegible] (2024) [illegible]
https://www.[illegible]

[illegible]

https://www.[illegible]

[illegible]

www.ingramcontent.com/pod-product-compliance
Lightning Source LLC
Chambersburg PA
CBHW071940210526
45479CB00002B/760
9798334964556